MEMOIRS
of the
American Mathematical Society

ISSN 0065-9266 (print)
ISSN 1947-6221 (online)

Number 1311

Local Boundedness, Maximum Principles, and Continuity of Solutions to Infinitely Degenerate Elliptic Equations with Rough Coefficients

Lyudmila Korobenko
Cristian Rios
Eric Sawyer
Ruipeng Shen

January 2021 • Volume 269 • Number 1311 (second of 7 numbers)

AMS AMERICAN MATHEMATICAL SOCIETY

Library of Congress Cataloging-in-Publication Data

Cataloging-in-Publication Data has been applied for by the AMS.
See http://www.loc.gov/publish/cip/.
DOI: https://doi.org/10.1090/memo/1311

Memoirs of the American Mathematical Society

This journal is devoted entirely to research in pure and applied mathematics.

Subscription information. Beginning with the January 2010 issue, *Memoirs* is accessible from `www.ams.org/journals`. The 2021 subscription begins with volume 269 and consists of six mailings, each containing one or more numbers. Subscription prices for 2021 are as follows: for paper delivery, US$1085 list, US$868 institutional member; for electronic delivery, US$955 list, US$764 institutional member. Upon request, subscribers to paper delivery of this journal are also entitled to receive electronic delivery. If ordering the paper version, add US$20 for delivery within the United States; US$80 for outside the United States. Subscription renewals are subject to late fees. See `www.ams.org/help-faq` for more journal subscription information. Each number may be ordered separately; *please specify number* when ordering an individual number.

Back number information. For back issues see `www.ams.org/backvols`.

Subscriptions and orders should be addressed to the American Mathematical Society, P. O. Box 845904, Boston, MA 02284-5904 USA. *All orders must be accompanied by payment.* Other correspondence should be addressed to 201 Charles Street, Providence, RI 02904-2213 USA.

Memoirs of the American Mathematical Society (ISSN 0065-9266 (print); 1947-6221 (online)) is published bimonthly (each volume consisting usually of more than one number) by the American Mathematical Society at 201 Charles Street, Providence, RI 02904-2213 USA. Periodicals postage paid at Providence, RI. Postmaster: Send address changes to Memoirs, American Mathematical Society, 201 Charles Street, Providence, RI 02904-2213 USA.

This publication is indexed in *Mathematical Reviews®, Zentralblatt MATH, Science Citation Index®, Science Citation Index™-Expanded, ISI Alerting Services℠, SciSearch®, Research Alert®, CompuMath Citation Index®, Current Contents®/Physical, Chemical & Earth Sciences.*
This publication is archived in *Portico* and *CLOCKSS*.
Printed in the United States of America.

∞ The paper used in this book is acid-free and falls within the guidelines established to ensure permanence and durability.
Visit the AMS home page at `https://www.ams.org/`

10 9 8 7 6 5 4 3 2 1   26 25 24 23 22 21

Contents

Preface vii

Part 1. Overview 1

Chapter 1. Introduction 5

Chapter 2. DeGiorgi Iteration, Local Boundedness, Maximum Principle and Continuity 9
1. Local boundedness and maximum principle for subsolutions 11
2. Continuity for weak solutions to the homogeneous equation 14

Chapter 3. Organization of the Proofs 17

Part 2. Abstract theory 19

Chapter 4. Local Boundedness 23
1. Orlciz norms 23
2. DeGiorgi iteration 26

Chapter 5. Maximum Principle 33

Chapter 6. Continuity 35

Part 3. Geometric theory 39

Chapter 7. Infinitely Degenerate Geometries 41
1. Calculation of the A-geodesics 41
2. Calculation of A-arc length 43
3. Integration over A-balls and Area 46
4. Higher dimensional geometries 61

Chapter 8. Orlicz Norm Sobolev Inequalities 67
1. Subrepresentation inequalities 67
2. $(1,1)$-Sobolev and $(1,1)$-Poincaré inequalities 84
3. Proportional vanishing L^1-Sobolev inequality 94
4. Orlicz Sobolev inequalities for bump functions Φ_N 96
5. Weak Sobolev inequality 102

Chapter 9. Geometric Theorems 105
1. Proofs of sufficiency 106
2. An unbounded weak solution 107

Part 4. Appendix 115

Appendix A. A Quasilinear Hypoellipticity Theorem 117

Appendix B. A Monge-Ampère Example 119

Appendix C. Absence of Radial Solutions 121

Bibliography 129

Abstract

We obtain local boundedness and maximum principles for weak subsolutions to certain infinitely degenerate elliptic divergence form inhomogeneous equations, and also continuity of weak solutions to homogeneous equations. For example, we consider the family $\{f_\sigma\}_{\sigma>0}$ with

$$f_\sigma(x) = e^{-\left(\frac{1}{|x|}\right)^\sigma}, \qquad -\infty < x < \infty,$$

of infinitely degenerate functions at the origin, and show that all weak solutions to the associated infinitely degenerate quasilinear equations of the form

$$\operatorname{div} A(x,u) \operatorname{grad} u = \phi(x), \quad A(x,z) \sim \begin{bmatrix} I_{n-1} & 0 \\ 0 & f(x_1)^2 \end{bmatrix},$$

with rough data A and ϕ, are locally bounded for admissible ϕ provided $0 < \sigma < 1$. We also show that these conditions are *necessary* for local boundedness in dimension $n \geq 3$, thus paralleling the known theory for the smooth Kusuoka-Strook operators $\frac{\partial^2}{\partial x_1^2} + \frac{\partial^2}{\partial x_2^2} + f_\sigma(x)^2 \frac{\partial^2}{\partial x_3^2}$. We also show that subsolutions satisfy a maximum principle under the same restriction on the degeneracy. Finally, continuity of solutions is derived in the homogeneous case $\phi \equiv 0$ under a more stringent assumption on the degeneracy, namely that $f \geq f_{3,\sigma}$ for $0 < \sigma < 1$ where

$$f_{3,\sigma}(x) = |x|^{\left(\ln\ln\ln \frac{1}{|x|}\right)^\sigma}, \qquad -\infty < x < \infty.$$

As an application we obtain weak hypoellipticity (i.e. smoothness of all weak solutions) of certain *infinitely* degenerate quasilinear equations in the plane

$$\frac{\partial u}{\partial x^2} + f(x, u(x,y))^2 \frac{\partial u}{\partial y^2} = 0,$$

with smooth data $f(x,z) \gtrsim f_{3,\sigma}(x)$ where $f(x,z)$ has a sufficiently mild nonlinearity and degeneracy.

In order to prove these theorems, we first establish abstract results in which certain Poincaré and Orlicz Sobolev inequalities are assumed to hold. We then develop subrepresentation inequalities for control geometries in order to obtain the needed Poincaré and Orlicz Sobolev inequalities.

Received by the editor July 25, 2017, and, in revised form, October 8, 2019.

Article electronically published on March 2, 2021.

DOI: https://doi.org/10.1090/memo/1311

2020 *Mathematics Subject Classification.* Primary 35B65, 35D30, 35H99; Secondary 51F99, 46E35.

Key words and phrases. Regularity, quasilinear equations, infinite degeneracy, rough coefficients, non-doubling control balls, subrepresentation inequality, Orlicz-Sobolev inequality.

Preface

There is a large and well-developed theory of elliptic and subelliptic equations with rough data, beginning with work of DeGiorgi-Nash-Moser, and also a smaller theory still in its infancy of infinitely degenerate elliptic equations with smooth data, beginning with work of Fedii and Kusuoka-Strook, and continued by Morimoto and Christ. Our purpose here is to initiate a study of the DeGiorgi regularity theory, as presented by Caffarelli-Vasseur, in the context of equations that are both infinitely degenerate elliptic and have rough data. This monograph can be viewed as taking the first steps in such an investigation and more specifically, in identifying a number of surprises encountered in the implementation of DeGiorgi iteration in the infinitely degenerate regime. The similar approach of Moser in the infinitely degenerate regime is initiated in our paper [**KoRiSaSh1**], but is both technially more complicated and more demanding of the underlying geometry. As a consequence, the results in [**KoRiSaSh1**] for local boundedness are considerably weaker than the sharp results obtained here with the DeGiorgi approach. On the other hand, the method of Moser does apply to obtain continuity for solutions to inhomogeneous equations, but at the expense of a much more elaborate proof strategy. The parallel approach of Nash seems difficult to adapt to the infinitely degenerate case, but remains a possibility for future research.

Part 1

Overview

The regularity theory of subelliptic linear equations with smooth coefficients is well established, as evidenced by the results of Hörmander [**Ho**] and Fefferman and Phong [**FePh**]. In [**Ho**], Hörmander obtained hypoellipticity of sums of squares of smooth vector fields whose Lie algebra spans at every point. In [**FePh**], Fefferman and Phong considered general nonnegative semidefinite smooth linear operators, and characterized subellipticity in terms of a containment condition involving Euclidean balls and "subunit" balls related to the geometry of the nonnegative semidefinite form associated to the operator.

The theory in the infinite regime however, has only had its surface scratched so far, as evidenced by the results of Fedii [**Fe**] and Kusuoka and Strook [**KuStr**]. In [**Fe**], Fedii proved that the two-dimensional operator $\frac{\partial}{\partial x^2}+f(x)^2\frac{\partial}{\partial y^2}$ is hypoelliptic merely under the assumption that f is smooth and positive away from $x=0$. In [**KuStr**], Kusuoka and Strook showed that under the same conditions on $f(x)$, the three-dimensional analogue $\frac{\partial^2}{\partial x^2}+\frac{\partial^2}{\partial y^2}+f(x)^2\frac{\partial^2}{\partial z^2}$ of Fedii's operator is hypoelliptic *if and only if* $\lim_{x\to 0} x\ln f(x)=0$. These results, together with some further refinements of Christ [**Chr**], illustrate the complexities associated with regularity in the infinite regime, and point to the fact that the theory here is still in its infancy.

The problem of extending these results to include quasilinear operators requires an understanding of the corresponding theory for linear operators with *nonsmooth* coefficients, generally as rough as the weak solution itself. In the elliptic case this theory is well-developed and appears for example in Gilbarg and Trudinger [**GiTr**] and many other sources. The key breakthrough there was the Hölder *apriori* estimate of DeGiorgi, and its later generalizations independently by Nash and Moser. The extension of the DeGiorgi-Nash-Moser theory to the subelliptic or finite type setting, originated in the work of Lanconelli [**Lan**] and a series of papers by Franchi and Lanconelli [**FrLan1**]-[**FrLan5**],[**Fr**], and then continued by many authors, including one of the present authors with Wheeden [**SaWh4**].

The subject of the present monograph[1] is the extension of DeGiorgi-Nash-Moser theory to the infinitely degenerate regime, and more specifically the techniques of DeGiorgi[2]. Our theorems fall into two broad categories. First, there is the *abstract theory* in all dimensions, in which we assume appropriate *Orlicz* Sobolev inequalities, as opposed to the familiar L^p Sobolev inequalities, and deduce local boundedness and maximum principles for weak subsolutions, and also continuity for weak solutions to homogeneous equations. This theory relies heavily on extensions of a lemma of DeGiorgi to the infinitely degenerate regime. Second, there is the *geometric theory*, in which we establish the required Orlicz Sobolev inequalities for large families of infinitely degenerate geometries, obtaining sharp results in dimension $n\geq 3$ for local boundedness. For this we need subrepresentation theorems with kernels of the form $K(x,y)=\frac{\widehat{d}(x,y)}{V(x,y)}$ where

$$\widehat{d}(x,y)=\min\left\{d(x,y),\frac{1}{|F'(x_1+d(x,y))|}\right\},$$

[1]This monograph is a distillation of the three papers [**KoRiSaSh1**], [**KoRiSaSh2**] and [**KoRiSaSh3**].

[2]We thank Pablo Raúl Stinga for bringing DeGiorgi's method to our attention.

$F = \ln \frac{1}{f}$ and $d(x, y)$ is the control distance associated with f, and where $V(x, y)$ is the volume of a control ball centered at x with radius $d(x, y)$. Typically, $\widehat{d}(x, y)$ is much smaller than $d(x, y)$ in the infinitely degenerate regime.

The results obtained here are of course also in their infancy, leaving many intriguing questions unanswered:

(1) In dimension $n = 2$ we have no counterexamples to continuity at all among the infinite order degeneracies that we consider.
(2) Our positive results here for continuity in any dimension apply only to homogeneous equations, due to the difficulty of implementing DeGiorgi iteration in the inhomogeneous case. On the other hand, Moser iteration does apply to inhomogeneous equations to obtain continuity of solutions, but requires a much more elaborate proof - see [**KoRiSaSh1**].

Finally, the contributions of Nash to the classical DeGiorgi-Nash-Moser theory revolve around moment estimates for solutions, and we have been unable to extend these to the infinitely degenerate regime, leaving a tantalizing loose end. We now turn to a more detailed description of these results and questions in the introduction that follows.

CHAPTER 1

Introduction

In 1971 Fedii proved in [**Fe**] that the linear second order partial differential operator

$$Lu(x,y) \equiv \left\{ \frac{\partial}{\partial x^2} + f(x)^2 \frac{\partial}{\partial y^2} \right\} u(x,y)$$

is *hypoelliptic*, i.e. every distribution solution $u \in \mathcal{D}'(\mathbb{R}^2)$ to the equation $Lu = \phi \in C^\infty(\mathbb{R}^2)$ in $\mathbb{R}^2$ is smooth, i.e. $u \in C^\infty(\mathbb{R}^2)$, provided:

- $f \in C^\infty(\mathbb{R})$,
- $f(0) = 0$ and f is positive on $(-\infty, 0) \cup (0, \infty)$.

The main feature of this remarkable theorem is that the order of vanishing of f at the origin is unrestricted, in particular it can vanish to infinite order. If we consider the analogous (special form) quasilinear operator,

$$\mathcal{L}u(x,y) \equiv \left\{ \frac{\partial}{\partial x^2} + f(x, u(x,y))^2 \frac{\partial}{\partial y^2} \right\} u(x,y),$$

then of course $f(x, u(x,y))$ makes no sense for u a distribution, but in the special case where $f(x,z) \approx f(x,0)$, the appropriate notion of hypoellipticity for $\mathcal{L}$ becomes that of $W_A^{1,2}(\mathbb{R}^2)$-hypoellipticity with $A \equiv \begin{bmatrix} 1 & 0 \\ 0 & f(x,0)^2 \end{bmatrix}$, which when A is understood, we refer to as simply *weak hypoellipticity*.

We say that $\mathcal{L}$ is $W_A^{1,2}(\mathbb{R}^2)$-hypoelliptic if every $W_A^{1,2}(\mathbb{R}^2)$-weak solution u to the equation $\mathcal{L}u = \phi$ is smooth for all smooth data $\phi(x,y)$. Here $u \in W_A^{1,2}(\mathbb{R}^2)$ is a $W_A^{1,2}(\mathbb{R}^2)$-weak solution to $L_{\text{quasi}} u = \phi$ if

$$-\int (\nabla w)^{\text{tr}} \begin{bmatrix} 1 & 0 \\ 0 & f(x, u(x,y))^2 \end{bmatrix} \nabla u = \int \phi w, \qquad \text{for all } w \in W_A^{1,2}(\mathbb{R}^2)_0 \, .$$

See below for a precise definition of the degenerate Sobolev space $W_A^{1,2}(\mathbb{R}^2)$, that informally consists of all $w \in L^2(\mathbb{R}^2)$ for which $\int (\nabla w)^{\text{tr}} A \nabla w < \infty$.

There is apparently no known $W_A^{1,2}(\mathbb{R}^2)$-hypoelliptic quasilinear operator $\mathcal{L}$ with coefficient $f(x,z)$ that vanishes to *infinite* order when $x = 0$, despite the abundance of results when f vanishes to *finite* order. However, in the infinite vanishing case, if we assume the stronger condition (1) below, which quantitatively limits the nonlinearity as $x \to 0$, and *in addition* condition (2) below,

(1) $\sup_{(x,z) \in (\mathbb{R} \setminus \{0\}) \times \mathbb{R}} \frac{|f(x,z) - f(x,0)|}{f(x,0)} \leq \frac{1}{2}$ and $\lim_{x \to 0} \sup_{z \in \mathbb{R}} \frac{\left|\frac{\partial f}{\partial z}(x,z)\right|}{f(x,0)} = 0$, and

(2) a $W_A^{1,2}(\mathbb{R}^2)$-weak solution u to the *homogeneous* equation $\mathcal{L}_{\text{quasi}} u = 0$ is *continuous*,

then in 2009 it was shown by Rios, Sawyer and Wheeden in [**RSaW2**] that for such solutions u we have $u \in C^\infty(\mathbb{R}^2)$. As a consequence of this result, together with

Theorem 21 below on continuity of weak solutions, we obtain that certain of these quasilinear operators $\mathcal{L}$ are $W_A^{1,2}(\mathbb{R}^2)$-hypoelliptic.

THEOREM 1. *Suppose that $f(x,z)$ is smooth in $\mathbb{R}^2$ and that in addition, $f(x,z)$ satisfies (1) and that for some $0<\sigma<1$, the function $f(x,0)$ satisfies*

$$c\,|x|^{\left(\ln\ln\ln\frac{1}{|x|}\right)^{\sigma}} \leq f(x,0), \qquad \text{for small } |x|>0. \tag{1.1}$$

Then every $W_A^{1,2}(\mathbb{R}^2)$-weak solution u to the homogeneous quasilinear equation $\mathcal{L}u=0$ is smooth.

See the appendix for a higher dimensional version of this theorem.

REMARK 2. The local sup norm bounds $\|D^\alpha u\|_{L^\infty}$ on the derivatives of u in Theorem 1 depend only on the constants c,σ in condition (1.1) and on the norm $\|u\|_{W_A^{1,2}(\mathbb{R}^2)}$ of the weak solution u in $W_A^{1,2}(\mathbb{R}^2)$.

Our method for proving regularity of weak solutions u to $\mathcal{L}u=\phi$ is to view u as a weak solution to the linear equation

$$Lu(x,y)\equiv\left\{\frac{\partial}{\partial x^2}+g(x,y)^2\frac{\partial}{\partial y^2}\right\}u(x,y)=\phi(x,y),$$

where both $g(x,y)=f(x,u(x,y))$ and $\phi(x,y)$ need no longer be smooth, but $g(x,y)$ satisfies the estimate

$$\frac{1}{C}f(x,0)\leq g(x,y)\leq Cf(x,0), \qquad x\in\mathbb{R},$$

and $\phi(x,y)$ is measurable and admissible - see below for definitions. The method we employ is an adaptation of DeGiorgi iteration. The infinite degeneracy of L forces our adaptation of DeGiorgi iteration to use Young functions in place of power functions.

Another motivation for this approach is the following three dimensional analogue of Fedii's equation, which Kusuoka and Strook [**KuStr**] considered in 1985

$$L_1\equiv\frac{\partial^2}{\partial x_1^2}+\frac{\partial^2}{\partial x_2^2}+f(x_1)^2\frac{\partial^2}{\partial x_3^2},$$

and showed the surprising result that when $f(x_1)$ is smooth and positive away from the origin, the smooth linear operator L_1 is hypoelliptic *if and only if*

$$\lim_{r\to 0} r\ln f(r)=0.$$

This is precisely the condition we show to be necessary and sufficient for local boundedness of weak solutions to our rough homogeneous equations. Thus we will begin with an abstract approach in higher dimensions, where we assume certain Orlicz Sobolev inequalities hold, and then specialize to geometries that are sufficient to prove the required Orlicz Sobolev inequalities.

More generally, we consider the divergence form equation

$$Lu=\nabla^{\mathrm{tr}}A(x)\nabla u=\phi, \qquad x\in\Omega,$$

and the corresponding second order special quasilinear equation (where only u, and not ∇u, appears nonlinearly),

$$\mathcal{L}u\equiv\nabla^{\mathrm{tr}}\mathcal{A}(x,u(x))\nabla u=\phi, \qquad x\in\Omega, \tag{1.2}$$

and we assume the following quadratic form condition on the 'quasilinear' matrix $\mathcal{A}(x,z)$,

$$c\,\xi^T A(x)\xi \le \xi^T \mathcal{A}(x,z)\xi \le C\,\xi^T A(x)\xi\ , \tag{1.3}$$

for a.e. $x \in \Omega$ and all $z \in \mathbb{R}$, $\xi \in \mathbb{R}^n$. Here c, C are positive constants and $A(x) = B(x)^{\mathrm{tr}} B(x)$ where $B(x)$ is a Lipschitz continuous $n \times n$ real-valued matrix defined for $x \in \Omega$. We define the A-gradient by

$$\nabla_A = B(x)\nabla\ , \tag{1.4}$$

and the associated degenerate Sobolev space $W_A^{1,2}(\Omega)$ to have norm

$$\|v\|_{W_A^{1,2}} \equiv \sqrt{\int_\Omega \left(|v|^2 + \nabla v^{\mathrm{tr}} A \nabla v\right)} = \sqrt{\int_\Omega \left(|v|^2 + |\nabla_A v|^2\right)}.$$

NOTATION 3. Somewhat informally, we use normal font $Lu = \nabla^{\mathrm{tr}} A(x) \nabla u$ for divergence form linear operators with nonnegative matrix $A(x)$, and we use calligraphic font $\mathcal{L}u = \nabla^{\mathrm{tr}} \mathcal{A}(x,u) \nabla u$ to denote the corresponding special form quasilinear operators with matrix $\mathcal{A}(x,u)$ comparable to $A(x)$.

In order to define the notion of weak solution to an inhomogeneous equation $\mathcal{L}u = \phi$ will assume that either $\phi \in L^2_{\mathrm{loc}}(\Omega)$ or that $\phi \in X(\Omega)$, where $X(\Omega)$ is the space of A-admissible functions defined as follows.

DEFINITION 4. Let Ω be a bounded domain in $\mathbb{R}^n$. Fix $x \in \Omega$ and $\rho > 0$. We say ϕ is *A-admissible* at (x,ρ) if

$$\|\phi\|_{X(B(x,\rho))} \equiv \sup_{v \in \left(W_A^{1,1}\right)_0 (B(x,\rho))} \frac{\int_{B(x,\rho)} |v\phi|\, dy}{\int_{B(x,\rho)} \|\nabla_A v\|\, dy} < \infty.$$

We say ϕ is *A-admissible* in Ω, written $\phi \in X(\Omega)$, if ϕ is *A-admissible* at (x,ρ) for all $B(x,\rho)$ contained in Ω.

DEFINITION 5. Let Ω be a bounded domain in $\mathbb{R}^n$ with A and $\mathcal{A}$ as above. Assume that $\phi \in L^2_{\mathrm{loc}}(\Omega) \cup X(\Omega)$. We say that $u \in W_A^{1,2}(\Omega)$ is a *weak solution* to $\mathcal{L}u = \phi$ provided

$$-\int_\Omega \nabla w(x)^{\mathrm{tr}} \mathcal{A}(x,u(x)) \nabla u = \int_\Omega \phi w$$

for all $w \in \left(W_A^{1,2}\right)_0(\Omega)$, where $\left(W_A^{1,2}\right)_0(\Omega)$ denotes the closure in $W_A^{1,2}(\Omega)$ of the subspace of Lipschitz continuous functions with compact support in Ω. Similarly, $u \in W_A^{1,2}(\Omega)$ is a *weak solution* to $Lu = \phi$ provided

$$-\int_\Omega \nabla w(x)^{\mathrm{tr}} A(x) \nabla u = \int_\Omega \phi w$$

for all $w \in \left(W_A^{1,2}\right)_0(\Omega)$.

Note that our structural condition (1.3) implies that the integral on the left above is absolutely convergent, and our assumption that $\phi \in L^2_{\mathrm{loc}}(\Omega) \cup X(\Omega)$

implies that the integral on the right above is absolutely convergent:

$$\begin{aligned} \int_{\Omega} |\phi w| &\leq \left(\int_{\Omega \cap \operatorname{supp} w} |\phi|^2 \right)^{\frac{1}{2}} \left(\int_{\Omega} |w|^2 \right)^{\frac{1}{2}}, \\ \int_{\Omega} |\phi w| &\leq \|\phi\|_{X(\Omega)} \int_{\Omega} \|\nabla_A w\| \leq \|\phi\|_{X(\Omega)} \left(\int_{\Omega} \|\nabla_A w\|^2 \right)^{\frac{1}{2}}. \end{aligned}$$

Weak sub and super solutions are defined by replacing $=$ with $\geq$ and $\leq$ respectively in the display above. We can define the gradient $\nabla_{\mathcal{A}}$ more generally for nonnegative semidefinite matrices $\mathcal{A}(x, u(x))$, where for convenience in notation we suppress the dependence on u.

DEFINITION 6. Given a real symmetric nonnegative semidefinite $n \times n$ matrix A, we can write $A = B^{\text{tr}} B$ with $B = D\left(\left\{\sqrt{\lambda_j}\right\}_{j=1}^n\right) P$ where $D\left(\{d_j\}_{j=1}^n\right) = \begin{bmatrix} d_1 & 0 & \cdots & 0 \\ 0 & d_2 & \cdots & 0 \\ \vdots & \vdots & \ddots & \vdots \\ 0 & 0 & \cdots & d_n \end{bmatrix}$ is the diagonal matrix having d_j along the diagonal, and P is the orthogonal matrix that diagonalizes A by the spectral theorem, i.e. $PAP^{-1} = D\left(\{\lambda_j\}_{j=1}^n\right)$. This representation is unique up to permutation of the eigenvalues $\{\lambda_j\}_{j=1}^n$ of A. Then we define $\nabla_A = B\nabla$. In the case that $\mathcal{A}(x) = A(x, u(x))$, this gives us a definition of $\nabla_{\mathcal{A}}$ for which

$$\int_{\Omega} (\nabla w)^{\text{tr}} \mathcal{A}(x, u(x)) \nabla w = \int_{\Omega} (\nabla w)^{\text{tr}} \mathcal{A}(x) \nabla w = \int_{\Omega} |\nabla_{\mathcal{A}} w|^2 .$$

We will first obtain *abstract* local boundedness, maximum principles, and continuity results, in which we *assume* appropriate Orlicz-Sobolev inequalities hold. Then we will apply our study of degenerate geometries to prove that these Orlicz-Sobolev inequalities hold in specific situations, thereby obtaining our *geometric* local boundedness, maximum principles, and continuity, in which we only assume information on the *size* of the degenerate geometries. The techniques used for local boundedness of weak subsolutions and maximum principles are very similar and so will be considered simultaneously. The proof of continuity builds on the local boundedness result with some further geometric properties, and thus will be treated separately.

CHAPTER 2

DeGiorgi Iteration, Local Boundedness, Maximum Principle and Continuity

Recall that the methods of DeGiorgi and Moser iteration play off a Sobolev inequality, that holds for all compactly supported functions, against a Cacciopoli inequality, that holds only for subsolutions or supersolutions of the equation. First, from results of Korobenko, Maldonado and Rios in [**KoMaRi**], it is known that if there exists a Sobolev bump inequality of the form

$$\|u\|_{L^q(\mu_B)} \leq Cr(B) \||\nabla_A u|\|_{L^p(\mu_B)}, \qquad u \in Lip_{\text{compact}}(B),$$

for some pair of exponents $1 \leq p < q \leq \infty$, and where the balls B are the Carnot-Carathéodory control balls for the degenerate vector field $\nabla_A = \left(\frac{\partial}{\partial x}, f\frac{\partial}{\partial y}\right)$ with radius $r(B)$, and $d\mu_B(x,y) = \frac{dxdy}{|B|}$ is normalized Lebesgue measure on B, then Lebesgue measure must be *doubling* on control balls. As a consequence, the function f *cannot* vanish to infinite order. Thus in order to have any hope of implementing either DeGiorgi or Moser iteration in the infinitely degenerate regime, we must search for a weaker Sobolev bump inequality, and the natural setting for this is an Orlicz Sobolev bump inequality

$$\|w\|_{L^\Phi(\mu_B)} \leq C\varphi(r(B)) \|\nabla_A w\|_{L^1(\mu_B)}, \qquad w \in Lip_{\text{compact}}(B),$$

where the Young function $\Phi(t)$ is increasing to ∞ and convex on $(0,\infty)$, but asymptotically closer to the identity t than any power function $t^{1+\sigma}$, $\sigma > 0$. The 'super-radius' $\varphi(r)$ here is nondecreasing and $\varphi(r) \geq r$.

We now recall the definition of Orlicz norms. Suppose that μ is a σ-finite measure on a set X, and $\Phi : [0,\infty) \to [0,\infty)$ is a Young function, which for our purposes is a convex piecewise differentiable (meaning there are at most finitely many points where the derivative of Φ may fail to exist, but right and left hand derivatives exist everywhere) function such that $\Phi(0) = 0$ and

$$\frac{\Phi(x)}{x} \to \infty \text{ as } x \to \infty.$$

Let L^Φ_* be the set of measurable functions $f : X \to \mathbb{R}$ such that the integral

$$\int_X \Phi(|f|)\, d\mu,$$

is finite, where as usual, functions that agree almost everywhere are identified. Since the set L^Φ_* may not be closed under scalar multiplication, we define L^Φ to be the linear span of L_*^Φ, and then define

$$\|f\|_{L^\Phi(\mu)} \equiv \inf\left\{k \in (0,\infty) : \int_X \Phi\left(\frac{|f|}{k}\right) d\mu \leq 1\right\}.$$

The Banach space $L^\Phi(\mu)$ is precisely the space of measurable functions f for which the norm $\|f\|_{L^\Phi(\mu)}$ is finite. The conjugate Young function $\widetilde{\Phi}$ is defined by $\widetilde{\Phi}' = (\Phi')^{-1}$ and can be used to give an equivalent norm

$$\|f\|_{L_*^\Phi(\mu)} \equiv \sup\left\{\int_X |fg|\, d\mu : \int_X \widetilde{\Phi}(|g|)\, d\mu \leq 1\right\}.$$

The above considerations motivate our approach, to which we now turn.

Let Ω be a bounded domain in $\mathbb{R}^n$. There is a quadruple $(\mathcal{A}, d, \varphi, \Phi)$ of objects of interest in our abstract local boundedness, maximum principle, and continuity theorems in Ω, namely

(1) the matrix $\mathcal{A} = \mathcal{A}(x,z)$ associated with our equation and the A-gradient,
(2) a Young function Φ appearing in our Orlicz Sobolev inequality,
(3) a metric d giving rise to the balls $B(x,r)$ that appear in our Orlicz Sobolev inequality, and also in our sequence $\{\psi_j\}_{j=1}^\infty$ of accumulating Lipschitz functions, and
(4) a positive function $\varphi(r)$ for $r \in (0,R)$ that appears in place of the radius r in our Orlicz Sobolev inequality.

For the abstract theory we will assume three connections between these objects, namely

- the existence of an appropriate sequence $\{\psi_j\}_{j=1}^\infty$ of accumulating Lipschitz functions that connects two of the objects of interest $\mathcal{A}$ and d, and
- an Orlicz Sobolev bump inequality,

$$\|w\|_{L^\Phi(B)} \leq \varphi(r(B))\, \|\nabla_A w\|_{L^1(B)}, \qquad \text{supp} w \subset B,$$

  that connects all four objects of interest $\mathcal{A}$, d, φ and Φ, and
- a proportional vanishing L^1-Sobolev inequality that is needed to prove continuity, namely

$$\int_B |u| \leq Cr(B)\frac{|B|}{|E|}\int_B |\nabla_A u|,$$

  for all $u \in Lip_1(B)$ that vanish on a subset E of a ball B.

We now describe these matters in more detail.

Definition 7 (Standard sequence of accumulating Lipschitz functions). Let Ω be a bounded domain in $\mathbb{R}^n$. Fix $r > 0$ and $x \in \Omega$. We define an $(\mathcal{A}, d)$-*standard* sequence of Lipschitz cutoff functions $\{\psi_j\}_{j=1}^\infty$ at (x,r), along with sets $B(x,r_j) \supset \text{supp}\psi_j$, to be a sequence satisfying $\psi_j = 1$on $B(x,r_{j+1})$, $r_1 = r$, $r_\infty \equiv \lim_{j\to\infty} r_j = \frac{1}{2}r$, $r_j - r_{j+1} = \frac{c}{j^\gamma}r$ for a uniquely determined constant c and $\gamma > 1$, and $\|\nabla_A \psi_j\|_\infty \lesssim \frac{j^\gamma}{r}$ with ∇_A as in (1.4) (see e.g. [**SaWh4**]).

We will need to assume the following single scale (Φ,φ)-Orlicz Sobolev bump inequality:

DEFINITION 8. Let Ω be a bounded domain in $\mathbb{R}^n$. Fix $x \in \Omega$ and $\rho > 0$. Then the single scale (Φ, φ)-Orlicz Sobolev bump inequality at (x, ρ) is:

$$\|w\|_{L^{\Phi}(B)} \leq \varphi(\rho) \|\nabla_A w\|_{L^1(B)}, \qquad w \in Lip_0(B(x,\rho)), \tag{2.1}$$

A particular family of Orlicz bump functions that is crucial for our theorem is the family

$$\Phi_N(t) \equiv \begin{cases} t(\ln t)^N & \text{if} \quad t \geq E = E_N = e^{2N} \\ (\ln E)^N t & \text{if} \quad 0 \leq t \leq E = E_N = e^{2N} \end{cases} .$$

The bump Φ_N is submultiplicative for each $N \geq 1$, i.e.

$$\Phi_N(st) \leq \Phi_N(s) \Phi_N(t), \qquad s, t > 0,$$

which can be easily seen using that for $s, t \geq e^{2N}$,

$$\begin{aligned} st[\ln(st)]^N &= st[\ln s + \ln t]^N \leq s[\ln s]^N t[\ln t]^N \\ &\iff \ln s + \ln t \leq [\ln s][\ln t], \end{aligned}$$

and then that $a + b \leq ab$ if $a, b \geq 2$. Submultiplicativity plays a critical role in proving our Orlicz Sobolev inequalities below.

For the inhomogeneous equation $\mathcal{L}u = \phi$ we will assume the forcing function ϕ is A-admissible in Ω. Finally, in order to deduce continuity of solutions to the homogeneous equation $\mathcal{L}u = 0$, we will need to assume a proportional vanishing L^1-Sobolev inequality.

DEFINITION 9. Let Ω be a bounded domain in $\mathbb{R}^n$. Then the proportional vanishing L^1-Sobolev inequality in Ω is:

$$\int_B |u| \leq Cr(B) \frac{|B|}{|E|} \int_{2B} |\nabla_A u|, \qquad \text{for all } u \in Lip_1(2B) \tag{2.2}$$

$$\text{that vanish on a subset } E \text{ of a ball } B \subset \Omega \text{ with } |E| \geq \frac{1}{2}|B|.$$

Note that the ball $2B$ on the right hand side is doubled. We are unable to prove this inequality with integration just over B on the right hand side.

1. Local boundedness and maximum principle for subsolutions

Recall that a measurable function u in Ω is *locally bounded above* at x if u can be modified on a set of measure zero so that the modified function $\widetilde{u}$ is bounded above in some neighbourhood of x.

THEOREM 10 (abstract local boundedness). *Let Ω be a bounded domain in $\mathbb{R}^n$ with $n \geq 2$. Suppose that $\mathcal{A}(x, z)$ is a nonnegative semidefinite matrix in $\Omega \times \mathbb{R}$ that satisfies the structural condition (1.3). Let $d(x, y)$ be a symmetric metric in Ω, and suppose that $B(x, r) = \{y \in \Omega : d(x, y) < r\}$ with $x \in \Omega$ are the corresponding metric balls. Fix $x \in \Omega$. Then every weak subsolution of (1.2) is* locally bounded above *at x provided there is $r_0 > 0$ such that:*

1. *the function ϕ is A-admissible at (x, r_0),*
2. *the single scale (Φ, φ)-Orlicz Sobolev bump inequality (2.1) holds at (x, r_0) with $\Phi = \Phi_N$ for some $N > 1$,*
3. *there exists an $(\mathcal{A}, d)$-*standard *accumulating sequence of Lipschitz cutoff functions at (x, r_0).*

REMARK 11. The hypotheses required for local boundedness of weak solutions to $Lu = \phi$ at a single *fixed* point x in Ω are quite weak; namely we only need that the inhomogeneous term ϕ is *A-admissible* at **just one** point (x, r_0) for some $r_0 > 0$, and that there are two *single scale* conditions relating the geometry to the equation at **the one** point (x, r_0).

REMARK 12. For the purposes of this paper we could simply take the metric d to be the Carnot-Carathéodory metric associated with A, but the present formulation allows for additional flexibility in the choice of balls used for DeGiorgi iteration in other situations.

In the special case that a weak subsolution u to (1.2) is *nonpositive* on the boundary of a ball $B(x, r_0)$, we can obtain a global boundedness inequality $\|u\|_{L^\infty(B(x,r_0))} \lesssim \|\phi\|_{X(B(x,r_0))}$ from the arguments used for Theorem 10, simply by noting that integration by parts no longer requires premultiplication by a Lipschitz cutoff function. Moreover, the ensuing arguments work just as well for an arbitrary bounded open set Ω in place of the ball $B(x, r_0)$, provided only that we assume our Sobolev inequality for Ω instead of for the ball $B(x, r_0)$. Of course there is no role played here by a superradius φ. This type of result is usually referred to as a *maximum principle*, and we now formulate our theorem precisely.

DEFINITION 13. Fix a bounded domain $\Omega \subset \mathbb{R}^n$. Then the Φ -Orlicz Sobolev bump inequality for Ω is:

$$\|w\|_{L^\Phi(\Omega)} \leq C \|\nabla_A w\|_{L^1(\Omega)}, \qquad w \in Lip_0(\Omega). \tag{2.3}$$

DEFINITION 14. Fix a bounded domain $\Omega \subset \mathbb{R}^n$. We say ϕ is *A-admissible* for Ω if

$$\|\phi\|_{X(\Omega)} \equiv \sup_{v \in \left(W_A^{1,1}\right)_0(\Omega)} \frac{\int_\Omega |v\phi| \, dy}{\int_\Omega \|\nabla_A v\| \, dy} < \infty.$$

We say a function $u \in W_A^{1,2}(\Omega)$ is *bounded by a constant* $\ell \in \mathbb{R}$ on the boundary $\partial\Omega$ if $(u-\ell)^+ = \max\{u-\ell, 0\} \in \left(W_A^{1,2}\right)_0(\Omega)$. We define $\sup_{x\in\partial\Omega} u(x)$ to be $\inf\left\{\ell \in \mathbb{R} : (u-\ell)^+ \in \left(W_A^{1,2}\right)_0(\Omega)\right\}$.

THEOREM 15 (abstract maximum principle). *Let Ω be a bounded domain in $\mathbb{R}^n$ with $n \geq 2$. Suppose that $\mathcal{A}(x,z)$ is a nonnegative semidefinite matrix in $\Omega \times \mathbb{R}$ that satisfies the structural condition (1.3). Let u be a nonnegative subsolution of (1.2). Then the following maximum principle holds,*

$$\operatorname{esssup}_{x\in\Omega} u(x) \leq \sup_{x\in\partial\Omega} u(x) + C \|\phi\|_{X(\Omega)},$$

where the constant C depends only on Ω, provided that:

1. *the function ϕ is A-admissible for Ω,*
2. *the Φ-Orlicz Sobolev bump inequality (2.3) for Ω holds with $\Phi = \Phi_N$ for some $N > 1$.*

In order to obtain a *geometric* local boundedness theorem, as well as a *geometric* maximum principle, we will take the metric d in Theorem 10 to be the Carnot-Caratheodory metric associated with the vector field ∇_A, and we will replace the hypotheses (2) and (3) in Theorem 10 with a geometric description of appropriate balls. For this we need to introduce a family of infinitely degenerate geometries that are simple enough that we can compute the balls, prove the required Sobolev Orlicz bump inequality, and define an appropriate accumulating sequence of Lipschitz cutoff functions.

We consider special quasilinear operators

$$\mathcal{L}u\left(x,y\right)\equiv\nabla^{\mathrm{tr}}\mathcal{A}\left(x,u\left(x\right)\right)\nabla u\left(x\right),\qquad x\in\Omega,$$

where $\Omega\subset\mathbb{R}^n$ and the $n\times n$ matrix $\mathcal{A}\left(x,z\right)$ has bounded measurable coefficients and is comparable to an n-dimensional matrix

$$A\left(x\right)\equiv\begin{bmatrix}1 & 0 & \cdots & 0 & 0\\ 0 & \ddots & & & \vdots\\ \vdots & & \ddots & 0 & 0\\ 0 & & 0 & 1 & 0\\ 0 & \cdots & 0 & 0 & f\left(x_1\right)^2\end{bmatrix},\qquad f\left(s\right)=e^{-F\left(s\right)},$$

by which we mean that

$$\begin{aligned}(2.4)\qquad & \frac{1}{C}\left(\xi_1^2+...+\xi_{n-1}^2+f\left(x_1\right)^2\xi_n^2\right)\\ \leq\ & \left(\xi_1,...,\xi_n\right)\mathcal{A}\left(x,z\right)\begin{pmatrix}\xi_1\\ \vdots\\ \xi_n\end{pmatrix}\\ \leq\ & C\left(\xi_1^2+...+\xi_{n-1}^2+f\left(x_1\right)^2\xi_n^2\right),\qquad a.e. x\in\Omega,\ z\in\mathbb{R}\text{ and }\xi\in\mathbb{R}^n,\end{aligned}$$

and where the degeneracy function $f\left(x\right)=e^{-F\left(x\right)}$ is even and there is $R>0$ such that F satisfies the following five structure conditions for some constants $C\geq 1$ and $\varepsilon>0$.

DEFINITION 16 (structure conditions). A function $F:(0,R)\rightarrow\mathbb{R}$ is said to satisfy *geometric structure conditions* if:

(1) $\lim_{x\to 0^+}F\left(x\right)=+\infty$;
(2) $F'\left(x\right)<0$ and $F''\left(x\right)>0$ for all $x\in(0,R)$;
(3) $\frac{1}{C}\left|F'\left(r\right)\right|\leq\left|F'\left(x\right)\right|\leq C\left|F'\left(r\right)\right|$ for $\frac{1}{2}r<x<2r<R$;
(4) $\frac{1}{-xF'(x)}$ is increasing in the interval $(0,R)$ and satisfies $\frac{1}{-xF'(x)}\leq\frac{1}{\varepsilon}$ for $x\in(0,R)$;
(5) $\frac{F''(x)}{-F'(x)}\approx\frac{1}{x}$ for $x\in(0,R)$.

REMARK 17. We make no smoothness assumption on f other than the existence of the second derivative f'' on the open interval $(0,R)$. Note also that at one extreme, f can be of finite type, namely $f\left(x\right)=x^{\alpha}$ for any $\alpha>0$, and at the other extreme, f can be of strongly degenerate type, namely $f\left(x\right)=e^{-\frac{1}{x^{\alpha}}}$ for any $\alpha>0$. Assumption (1) rules out the elliptic case $f\left(0\right)>0$.

Using these geometric structure conditions, we can show that standard sequences of Lipschitz cutoff functions always exist for our geometries.

LEMMA 18. *If $\gamma>1$ and $A\left(x\right)$ is a* continuous *nonnegative semidefinite $n\times n$ matrix valued function on a bounded domain $\Omega\subset\mathbb{R}^n$ as above, and if d is the associated control metric, then for every $r>0$ and $x\in\Omega$, there is an (A,d)-*standard *sequence of Lipschitz cutoff functions $\left\{\psi_j\right\}_{j=1}^{\infty}$ at (x,r), associated with balls $B(x,r_j)\supset$* suppψ_j *as in Definition 7.*

PROOF. This follows immediately from Proposition 68 on page 90 of [**SaWh4**], once we observe that in the proof of Proposition 68, we can take N to be any real number greater than 1 (so that $\sum_{j=1}^{\infty} j^{-N} < \infty$), and that the assumption of the containment condition in Proposition 68 there was *only* used in the proof to conclude that the annuli $B(x,t) \setminus B(x,s)$ have positive Euclidean thickness for $0 < s < t < \infty$ - i.e. that the boundaries $\partial B(x,t)$ and $\partial B(x,s)$ are pairwise disjoint. This is certainly the case for the control balls $B(x,r)$ associated with our geometries F satisfying Definition 16, and so the proof of Proposition 68 of [**SaWh4**] applies to prove Lemma 18. □

In the next theorem we will consider the geometry of balls defined by

$$F_\sigma(r) = r^{-\sigma},$$

where $\sigma > 0$. Note that $f_\sigma = e^{-F_\sigma}$ vanishes to infinite order at $r = 0$, and that f_σ vanishes to a faster order than $f_{\sigma'}$ if $\sigma > \sigma'$. We also define the simpler linear operator

$$Lu \equiv \operatorname{div} A(x) \nabla u.$$

with $A(x)$ as in (1.3).

THEOREM 19. *Suppose that $\Omega \subset \mathbb{R}^n$ is a domain in $\mathbb{R}^n$ with $n \geq 2$ and that*

$$\mathcal{L}u \equiv \operatorname{div}\mathcal{A}(x,u) \nabla u, \qquad x = (x_1, ..., x_n) \in \Omega,$$

where $\mathcal{A}(x,z) \sim \begin{bmatrix} I_{n-1} & 0 \\ 0 & f(x_1)^2 \end{bmatrix}$, *$I_{n-1}$ is the $(n-1) \times (n-1)$ identity matrix, $\mathcal{A}$ has bounded measurable components, and the geometry $F = -\ln f$ satisfies the geometric structure conditions in Definition* 16.

1. *If $F \leq F_\sigma$ for some $0 < \sigma < 1$, then every weak subsolution to $\mathcal{L}u = \phi$ with A-admissible ϕ is locally bounded in Ω.*
2. *On the other hand, if $n \geq 3$ and $\sigma \geq 1$, then there exists a locally unbounded weak solution u in a neighbourhood of the origin in $\mathbb{R}^n$ to the equation $Lu = 0$ with geometry $F = F_\sigma$.*

THEOREM 20. *Suppose that F satisfies the geometric structure conditions in Definition* 16 *and $F \leq F_\sigma$ for some $0 < \sigma < 1$. Assume that u is a weak subsolution to $\mathcal{L}u = \phi$ in a domain $\Omega \subset \mathbb{R}^n$ with $n \geq 2$, where $\mathcal{L}$ has degeneracy F and ϕ is A-admissible. Moreover, suppose that u is bounded in the weak sense on the boundary $\partial\Omega$. Then u is globally bounded in Ω and satisfies*

$$\sup_{\Omega} u \leq \sup_{\partial\Omega} u + C \|\phi\|_{X(\Omega)}$$

with the constant C depending only on Ω.

2. Continuity for weak solutions to the homogeneous equation

Now we consider weak solutions to the *homogeneous* equation $\mathcal{L}u = 0$ in a domain $\Omega \subset \mathbb{R}^n$, and improve our local boundedness conclusion to continuity. As before we first give an abstract theorem, and follow this with a geometric theorem. The abstract theorem requires our proportional vanishing L^1-Sobolev inequality as given in Definition 9 above.

THEOREM 21. *Suppose that $\Omega \subset \mathbb{R}^n$ is a domain in $\mathbb{R}^n$ with $n \geq 2$. Suppose the proportional vanishing Poincaré inequality (2.2) holds. For every ball $B_{r_0} = B(x, r_0)$ contained in Ω, assume that the local boundedness inequality,*

$$\|u_+\|_{L^\infty\left(B_{\frac{r}{2}}\right)} \leq \frac{1}{2\sqrt{\delta(r)}} \sqrt{\frac{1}{|B_{3r}|} \int_{B_r} u_+^2}\,, \tag{2.5}$$

holds for $\mathcal{L}u = 0$ in B_r with $\delta(r) > 0$, for all $0 < r < r_0$. Let $r_j = \frac{r_0}{4^j}$, and set $B_j \equiv B_{r_j} = B(x, r_j)$. Suppose furthermore that the following summability condition holds,

$$\sum_{j=1}^{\infty} \lambda_j = \infty,$$

where $\lambda_j \equiv \frac{1}{2^{3+\frac{C_3}{\delta^2(r_j)}}}$ and $B_j = B(0, r_j) = B\left(0, \frac{r_0}{4^j}\right)$, and where C_3 is a positive constant. Then if u is a weak solution to $\mathcal{L}u = 0$ in B_{r_0}, we have $u \in C(B_{r_0})$.

For the geometric continuity theorem we need to consider a less degenerate family of geometries. For $k \geq 0$ and $0 < \sigma < \infty$, define $F_{k,\sigma}(r) = \left(\ln \frac{1}{r}\right)\left(\ln^{(k)} \frac{1}{r}\right)^\sigma$ and $f_{k,\sigma}(r) = e^{-F_{k,\sigma}(r)} = r^{\left(\ln^{(k)} \frac{1}{r}\right)^\sigma}$. Note that $F_{0,\sigma}(r) = \left(\ln \frac{1}{r}\right) \frac{1}{r^\sigma}$ and $F_\sigma(r) = \frac{1}{r^\sigma}$ for $0 < \sigma < \infty$ are essentially the same families of geometries.

THEOREM 22. *Suppose that $\Omega \subset \mathbb{R}^n$ is a domain in $\mathbb{R}^n$ with $n \geq 2$ and that*

$$\mathcal{L}u \equiv \operatorname{div} \mathcal{A}(x, u) \nabla u, \qquad x = (x_1, ..., x_n) \in \Omega,$$

where $\mathcal{A}(x, z) \sim \begin{bmatrix} I_{n-1} & 0 \\ 0 & f(x_1)^2 \end{bmatrix}$, I_{n-1} is the $(n-1) \times (n-1)$ identity matrix, $\mathcal{A}$ has bounded measurable components, and the geometry $F = -\ln f$ satisfies the structure conditions in Definition 16.

1. *If $F \leq F_{3,\sigma}$ for some $0 < \sigma < 1$, then every weak solution to $\mathcal{L}u = 0$ is continuous in Ω.*
2. *On the other hand, if $n \geq 3$ and $\sigma \geq 1$, then there exists a locally unbounded weak solution u in a neighbourhood of the origin in $\mathbb{R}^n$ to the equation $Lu = 0$ with geometry $F = F_{0,\sigma}$.*

CHAPTER 3

Organization of the Proofs

In Part 2 we use DeGiorgi iteration to prove the abstract local boundedness, maximum principle and continuity theorems. Then in Part 3 we first calculate geodesics and volumes of balls in our geometries F satisfying the geometric structure conditions in Definition 16, and second establish subrepresentation and Orlicz Sobolev inequalities. Finally then we can prove the geometric theorems. The Appendix in Part 4 contains our higher dimensional quasilinear hypoellipticity theorem, and also contains two tangential results of a negative nature - the first involving the Monge-Ampère equation, and the second demonstrating the absence of nonconstant metric radial solutions to equations with radial right hand side.

Part 2

Abstract theory

There are three main ingredients needed to prove local boundedness, maximum principle and continuity results, namely the Orlicz Sobolev inequality for compactly supported functions, the proportional vanishing L^1 Sobolev inequality, and the Caccioppoli inequality for subsolutions of the degenerate equation. We start with the Orlicz Sobolev inequality we need.

Let Φ be a Young function on $(0,\infty)$ and let F be a geometry satisfying the geometric structure conditions in Definition 16. We will assume initially that we have an Orlicz Sobolev norm inequality for the control balls B in some domain $\Omega\subset\mathbb{R}^n$:

$$\|w\|_{L^{\Phi}(\mu_B)}\leq\varphi\left(r_0\right)\ \|\nabla_A w\|_{L^1(\mu_B)}, \qquad w\in\left(W_A^{1,1}\right)_0(B), \tag{3.1}$$

for some increasing 'superradius' function $\varphi\left(r_0\right)\geq r_0$, where r_0 is the control radius of a control ball B and μ_B is normalized Lebesgue measure on B. We prove this inequality for appropriate geometries and superradii below in Proposition 80 and Lemma 78.

Next, we establish a Caccioppoli inequality for weak subsolutions that holds independently of any geometric considerations.

PROPOSITION 23. *If u is a weak subsolution to $\mathcal{L}u=\phi$ with $\mathcal{L}$ as in (1.2), (1.3), and A-admissible ϕ, then the the following Caccioppoli inequality holds for u_+ on a ball B with Lipschitz ψ supported in B:*

$$\int|\nabla_A(\psi u_+)|^2d\mu\leq C\left(\|\psi\|_{L^2(\mu)}^2\|\phi\|_X^2+\|\nabla_A\psi\|_{L^\infty}^2\int_B u_+^2d\mu\right), \tag{3.2}$$

where $d\mu=\frac{dx}{|B|}$, and where the constant C depends only on the constants in (1.3) and not on $r(B)$.

PROOF. Recall we use C and c to denote constants that may change from line to line. For convenience in notation we denote the matrix function $\mathcal{A}\left(x,u\left(x\right)\right)$ by $\mathcal{A}$, and its associated gradient by $\nabla_{\mathcal{A}}$. Since our conclusion involves the matrix A, while the definition of u being a subsolution involves the matrix $\mathcal{A}$, we must apply care in using the comparability of the matrices A and $\mathcal{A}$ in the positive definite sense. We have the pointwise vector identity

$$\begin{aligned}
&|\nabla_{\mathcal{A}}(\psi u_+)|^2-\nabla\left(\psi^2u_+\right)\mathcal{A}\nabla u_+\\
=\;&|\psi\nabla_{\mathcal{A}}u_++u_+\nabla_{\mathcal{A}}\psi|^2-\left[\psi^2\nabla u_++u_+\nabla\psi^2\right]\mathcal{A}\nabla u_+\\
=\;&\left|\psi\nabla_{\mathcal{A}}u_+\right|^2+2\left\langle\psi\nabla_{\mathcal{A}}u_+,u_+\nabla_{\mathcal{A}}\psi\right\rangle+\left|u_+\nabla_{\mathcal{A}}\psi\right|^2\\
&-\left|\psi\nabla_{\mathcal{A}}u_+\right|^2-2\left\langle u_+\nabla_{\mathcal{A}}\psi,\psi\nabla_{\mathcal{A}}u_+\right\rangle\\
=\;&\left|u_+\nabla_{\mathcal{A}}\psi\right|^2.
\end{aligned}$$

Then using (1.3), we obtain

$$c|\nabla_A(\psi u_+)|^2\leq\nabla\left(\psi^2u_+\right)\mathcal{A}\nabla u_++C\left|u_+\nabla_A\psi\right|^2,$$

and using the fact that $d\mu\left(x\right)$ is a constant multiple of Lebesgue measure dx, we integrate to obtain that

$$\|\nabla_A(\psi u_+)\|_{L^2(\mu)}^2\leq C\int\nabla\left(\psi^2u_+\right)\mathcal{A}\nabla u_+d\mu+C\left\|u_+\nabla_A\psi\right\|_{L^2(\mu)}^2. \tag{3.3}$$

Next, since u is a weak subsolution to $\mathcal{L}u = \phi$, we have $\nabla \mathcal{A} \nabla u \geq \phi$ in the weak sense, which implies

$$\begin{aligned}\int \nabla\left(\psi^2 u_+\right) \mathcal{A} \nabla u_+ d\mu &= \int \nabla\left(\psi^2 u_+\right) \mathcal{A} \nabla u d\mu \\ &\leq -\int\left(\psi^2 u_+\right) \phi d\mu \\ &\leq \|\phi\|_X \int\left|\nabla_{\mathcal{A}}\left(\psi^2 u_+\right)\right| d\mu .\end{aligned} \tag{3.4}$$

Now we compute that

$$\begin{aligned}\int\left|\nabla_{\mathcal{A}}\left(\psi^2 u_+\right)\right| d\mu &= \int\left|\psi u_+ \nabla_{\mathcal{A}} \psi+\psi \nabla_{\mathcal{A}}(\psi u_+)\right| d\mu \\ &\leq \|\psi u_+\|_{L^2(\mu)}\|\nabla_{\mathcal{A}} \psi\|_{L^2(\mu)}+\|\nabla_{\mathcal{A}}(\psi u_+)\|_{L^2(\mu)}\|\psi\|_{L^2(\mu)} \\ &\leq \|\nabla_{\mathcal{A}} \psi\|_{L^2(d\mu)}\|\psi u_+\|_{L^2(\mu)}+\|\psi\|_{L^2(d\mu)}\|\nabla_{\mathcal{A}}(\psi u_+)\|_{L^2(\mu)},\end{aligned}$$

and combining this with (3.3) and (3.4), and using (1.3) again, we obtain

$$\begin{aligned}\|\nabla_A(\psi u_+)\|_{L^2(\mu)}^2 \leq\ & C\|\phi\|_X\left\{\|\nabla_A \psi\|_{L^2(d\mu)}\|\psi u_+\|_{L^2(\mu)}\right. \\ &\left.+\|\psi\|_{L^2(d\mu)}\|\nabla_A(\psi u_+)\|_{L^2(\mu)}\right\} \\ &+C\|\nabla_A \psi\|_{L^\infty}^2 \int_B u_+^2 d\mu .\end{aligned} \tag{3.5}$$

To estimate the first term on the right hand side of (3.5), we apply Young's inequality twice to get

$$\begin{aligned}&\|\phi\|_X\left(\|\nabla_A \psi\|_{L^2(d\mu)}\|\psi u_+\|_{L^2(\mu)}+\|\psi\|_{L^2(d\mu)}\|\nabla_A(\psi u_+)\|_{L^2(\mu)}\right) \\ \leq\ & \frac{1}{\varepsilon}\|\phi\|_X^2+\varepsilon\left(\|\nabla_A \psi\|_{L^2(d\mu)}\|\psi u_+\|_{L^2(\mu)}+\|\psi\|_{L^2(d\mu)}\|\nabla_A(\psi u_+)\|_{L^2(\mu)}\right)^2 \\ \leq\ & \frac{1}{\varepsilon}\|\phi\|_X^2+C\varepsilon\left(\|\nabla_A \psi\|_{L^2(d\mu)}^2\|\psi u_+\|_{L^2(\mu)}^2+\|\psi\|_{L^2(d\mu)}^2\|\nabla_A(\psi u_+)\|_{L^2(\mu)}^2\right),\end{aligned}$$

and combining this with (3.5), we obtain

$$\begin{aligned}\|\nabla_A(\psi u_+)\|_{L^2(\mu)}^2 \leq & C\frac{1}{\varepsilon}\|\phi\|_X^2+C\varepsilon\|\nabla_A \psi\|_{L^2(d\mu)}^2\|\psi u_+\|_{L^2(\mu)}^2 \\ &+C\varepsilon\|\psi\|_{L^2(d\mu)}^2\|\nabla_A(\psi u_+)\|_{L^2(\mu)}^2+C\|\nabla_A \psi\|_{L^\infty}^2 \int_B u_+^2 d\mu .\end{aligned} \tag{3.6}$$

Now choose ε so small that

$$C\varepsilon\|\psi\|_{L^2(d\mu)}^2=C\varepsilon \frac{1}{|B|} \int_B \psi^2 \leq C\varepsilon\|\psi\|_{L^\infty}^2<\frac{1}{2},$$

and then absorb the third term on the right hand side of (3.6) into its left hand side to obtain

$$\|\nabla_A(\psi u_+)\|_{L^2(\mu)}^2 \leq C\left(\|\psi\|_{L^2(d\mu)}^2\|\phi\|_X^2+\|\nabla_A \psi\|_{L^\infty}^2 \int_B u_+^2 d\mu\right)$$

upon using $\|\nabla_A \psi\|_{L^2(d\mu)}^2 \leq \|\nabla_A \psi\|_{L^\infty}^2$. This completes the proof of (3.2). ∎

REMARK 24. It is important to note that (3.2) holds for u_+ whenever u is a weak subsolution *without* assuming that u_+ is also a subsolution.

COROLLARY 25 (of the proof). *Let u be a weak subsolution to $\mathcal{L}u = \phi$ with $\mathcal{L}$ as in (1.2), (1.3), and A-admissible ϕ, and suppose for some ball B, $P > 0$, and a nonnegative function $v \in W_A^{1,2}$ there holds*

$$\|\phi\|_X^2 \leq Pv(x), \quad a.e.\ x \in \{u > 0\} \cap B. \tag{3.7}$$

Then

$$\int |\nabla_A(\psi u_+)|^2 d\mu \leq C\left(\|\nabla_A \psi\|_{L^\infty} + P\right)^2 \int \left(u_+^2 + v^2\right) d\mu, \tag{3.8}$$

where $d\mu = \frac{dx}{|B|}$, and where ψ is a Lipschitz cutoff function supported in B.

PROOF. First, recall that from (3.3) we have

$$\int |\nabla_A(\psi u_+)|^2 d\mu \leq C \int \nabla\left(\psi^2 u_+\right) \mathcal{A}\nabla u_+ d\mu + C\|\nabla_A \psi\|_{L^\infty}^2 \int_B u_+^2 d\mu\ , \tag{3.9}$$

where the constant C depends only on constants in (1.3). Now using (3.4) and (3.7) we have

$$\begin{aligned}
&\int \nabla\left(\psi^2 u_+\right) \mathcal{A}\nabla u_+ d\mu \\
&\leq \|\phi\|_X \int \left|\nabla_{\mathcal{A}}\left(\psi^2 u_+\right)\right| d\mu \leq P \int v \left|\nabla_{\mathcal{A}}\left(\psi^2 u_+\right)\right| d\mu \\
&= P \int \left|\psi u_+ v \nabla_{\mathcal{A}} \psi d\mu + \psi v \nabla_{\mathcal{A}}(\psi u_+)\right| d\mu \\
&\leq P\|\psi u_+\|_{L^2(\mu)} \|v\nabla_{\mathcal{A}}\psi\|_{L^2(\mu)} + \varepsilon\|\nabla_{\mathcal{A}}(\psi u_+)\|_{L^2(\mu)}^2 + \frac{P^2}{\varepsilon}\|\psi v\|_{L^2(\mu)}^2 \\
&\leq CP\|\psi u_+\|_{L^2(\mu)} \|v\nabla_A\psi\|_{L^2(\mu)} + \varepsilon C\|\nabla_A(\psi u_+)\|_{L^2(\mu)}^2 + \frac{P^2}{\varepsilon}\|\psi v\|_{L^2(\mu)}^2 \\
&\leq CP\|\nabla_A\psi\|_{L^\infty} \int \left(u_+^2 + v^2\right) d\mu + C\varepsilon\|\nabla_A(\psi u_+)\|_{L^2(\mu)}^2 + \frac{P^2}{\varepsilon}\int (\psi v)^2 d\mu,
\end{aligned}$$

where in passing from the third line to the fourth line above, we have replaced $\mathcal{A}$ with A at the expense of multiplying by the constant C in (1.3). Combining this with (3.9), and choosing ε small enough to absorb the second summand on the right, we obtain

$$\int |\nabla_A(\psi u_+)|^2 d\mu \leq C\left(\|\nabla_A \psi\|_{L^\infty} + P\right)^2 \int \left(u_+^2 + v^2\right) d\mu.$$

□

CHAPTER 4

Local Boundedness

We begin with a short review of that part of the theory of Orlicz norms that is relevant for us.

1. Orlciz norms

Recall that if $\Phi : [0,\infty) \to [0,\infty)$ is a Young function, we define L^{Φ} to be the linear span of L_*^{Φ}, the set of measurable functions $f : X \to \mathbb{R}$ such that the integral $\int_X \Phi(|f|)\, d\mu$ is finite, and then define

$$\|f\|_{L^{\Phi}(\mu)} \equiv \inf\left\{k \in (0,\infty) : \int_X \Phi\left(\frac{|f|}{k}\right) d\mu \leq 1\right\}.$$

In our application to DeGiorgi iteration the convex bump function $\Phi(t)$ will satisfy in addition:

- The function $\frac{\Phi(t)}{t}$ is positive, nondecreasing and tends to ∞ as $t \to \infty$;
- Φ is submultiplicative on an interval (E,∞) for some $E > 1$:

$$\Phi(ab) \leq \Phi(a)\Phi(b), \qquad a,b > E. \tag{4.1}$$

Note that if we consider more generally the quasi-submultiplicative condition,

$$\Phi(ab) \leq K\Phi(a)\Phi(b), \qquad a,b > E, \tag{4.2}$$

for some constant K, then $\Phi(t)$ satisfies (4.2) if and only if $\Phi_K(t) \equiv K\Phi(t)$ satisfies (4.1). Thus we can alway rescale a quasi-submultiplicative function to be submultiplicative.

Now let us consider the *linear extension* Φ_{ext} of a function $\Phi : [E,\infty) \to [0,\infty)$ to the entire positive real axis $(0,\infty)$ given by

$$\Phi_{\text{ext}}(t) = \begin{cases} \Phi(t) & \text{if} \quad E \leq t < \infty \\ \frac{\Phi(E)}{E}t & \text{if} \quad 0 \leq t \leq E \end{cases}.$$

We claim that Φ_{ext} is submultiplicative on $(0,\infty)$, i.e.

$$\Phi_{\text{ext}}(ab) \leq \Phi_{\text{ext}}(a)\Phi_{\text{ext}}(b), \qquad a,b > 0.$$

In fact, the identity $\frac{\Phi_{\text{ext}}(t)}{t} = \frac{\Phi_{\text{ext}}(\max\{t,E\})}{\max\{t,E\}}$ and the monotonicity of $\frac{\Phi(t)}{t}$ imply

$$\begin{aligned} \frac{\Phi_{\text{ext}}(ab)}{ab} &\leq \frac{\Phi_{\text{ext}}(\max\{a,E\}\max\{b,E\})}{\max\{a,E\}\max\{b,E\}} \\ &\leq \frac{\Phi_{\text{ext}}(\max\{a,E\})}{\max\{a,E\}} \cdot \frac{\Phi_{\text{ext}}(\max\{b,E\})}{\max\{b,E\}} = \frac{\Phi_{\text{ext}}(a)}{a}\frac{\Phi_{\text{ext}}(b)}{b}. \end{aligned}$$

CONCLUSION 26. If $\Phi : [E,\infty) \to \mathbb{R}^+$ is a submultiplicative piecewise differentiable strictly convex function with the property that $\frac{\Phi(t)}{t}$ is nondecreasing on

$[E,\infty)$, then we can extend Φ to a submultiplicative piecewise differentiable convex function Φ_{ext} on $[0,\infty)$ that vanishes at 0 *if and only if*

$$\Phi'(E) \geq \frac{\Phi(E)}{E}. \tag{4.3}$$

So now we suppose that Φ and $E \in (0,\infty)$ satisfy (4.3) and that Φ_{ext} is a submultiplicative piecewise differentiable convex function on $[0,\infty)$ that vanishes at 0, and moreover is strictly convex on (E,∞). Let

$$\Psi(t) = \Phi'_{\text{ext}}(t) = \begin{cases} \Phi'(t) & \text{if} \quad E \leq t < \infty \\ \frac{\Phi(E)}{E} & \text{if} \quad 0 \leq t < E \end{cases}.$$

Now Ψ is increasing on $(0,\infty)$, but is constant on $(0,E)$, and in addition has a jump discontinuity at E if $\Phi'(E) > \frac{\Phi(E)}{E}$. Since $\Psi^{(-1)}$ does not exist on all of $(0,\infty)$, we instead **define** the function $\Psi^{(-1)}$ by reflecting the graph of Ψ about the line $s=t$ in the (t,s)-plane:

$$\Psi^{(-1)}(s) = \begin{cases} (\Phi')^{(-1)}(s) & \text{if} \quad \Phi'(E) \leq s < \infty \\ E & \text{if} \quad \frac{\Phi(E)}{E} \leq s < \Phi'(E) \\ 0 & \text{if} \quad 0 < s < \frac{\Phi(E)}{E} \end{cases}.$$

Finally we **define** the conjugate function $\widetilde{\Phi}$ of Φ by the formula

$$\begin{aligned} \widetilde{\Phi}(s) &\equiv \int_0^s \Psi^{(-1)}(x)\,dx \\ &= \begin{cases} E\Phi'(E) - \Phi(E) + \int_{\Phi'(E)}^s (\Phi')^{(-1)}(x)\,dx & \text{if} \quad \Phi'(E) \leq s < \infty \\ Es - \Phi(E) & \text{if} \quad \frac{\Phi(E)}{E} \leq s < \Phi'(E) \\ 0 & \text{if} \quad 0 < s < \frac{\Phi(E)}{E} \end{cases}. \end{aligned}$$

One now has the standard Young's inequality for the pair of functions $\left(\Psi,\Psi^{(-1)}\right)$:

$$ts \leq \Phi(t) + \widetilde{\Phi}(s), \qquad 0 \leq t,s < \infty.$$

Indeed, the left hand side is the area of the rectangle $[0,t]\times[0,s]$ in the (t,s)-plane, which is at most the area $\int_0^t \Psi(y)\,dy$ under the graph of Ψ up to t plus the area $\int_0^s \Psi^{(-1)}(y)\,dy$ under the graph of $\Psi^{(-1)}$ up to s. As a consequence we have the following generalization Hölder's inequality:

$$\begin{aligned} \frac{\int_X |f(x)g(x)|\,d\mu(x)}{\|f\|_{L^\Phi(\mu)}\|g\|_{L^{\widetilde{\Phi}}(\mu)}} &= \int_X \frac{|f(x)|}{\|f\|_{L^\Phi(\mu)}}\frac{|g(x)|}{\|g\|_{L^{\widetilde{\Phi}}(\mu)}}d\mu(x) \\ &\leq \int_X \left\{\Phi\left(\frac{|f(x)|}{\|f\|_{L^\Phi(\mu)}}\right) + \widetilde{\Phi}\left(\frac{|g(x)|}{\|g\|_{L^{\widetilde{\Phi}}(\mu)}}\right)\right\}d\mu(x) \leq 1+1=2. \end{aligned}$$

We now restrict attention to a particular family of bump functions Φ_N, that we will use in our adaptation of the DeGiorgi iteration scheme, namely

$$\Phi_N(t) \equiv \begin{cases} t(\ln t)^N & \text{if} \quad t \geq E = E_N = e^{2N} \\ (\ln E)^N t & \text{if} \quad 0 \leq t \leq E = E_N = e^{2N} \end{cases}. \tag{4.4}$$

We have that Φ_N is submultiplicative for $s,t \geq e^{2N}$ and $N \geq 1$ upon using that

$$\begin{aligned} st[\ln(st)]^N &= st[\ln s + \ln t]^N \leq s[\ln s]^N t[\ln t]^N \\ &\iff \ln s + \ln t \leq [\ln s][\ln t], \end{aligned}$$

and $a+b \leq ab$ if $a, b \geq 2$. Thus the above considerations apply to show that for $N \geq 1$, the Young function $\Phi = \Phi_N$ is convex and submultiplicative, and that the Hölder inequality

$$\int_X |f(x)\, g(x)|\, d\mu(x) \leq 2\, \|f\|_{L^\Phi(\mu)}\, \|g\|_{L^{\widetilde{\Phi}}(\mu)}$$

holds with conjugate function

$$\widetilde{\Phi}(s) = \begin{cases} \frac{1}{2}\left(2Ne^2\right)^N + \int_{\frac{3}{2}(2N)^N}^s (\Phi')^{(-1)}(x)\, dx & \text{if} \quad \frac{3}{2}(2N)^N \leq s < \infty \\ e^{2N}\left\{s - (2N)^N\right\} & \text{if} \quad (2N)^N \leq s < \frac{3}{2}(2N)^N \\ 0 & \text{if} \quad 0 < s < (2N)^N \end{cases},$$

since $\frac{d}{dt}\left\{t(\ln t)^N\right\} = (\ln t)^N + N(\ln t)^{N-1} = (\ln t)^N\left(1 + \frac{N}{\ln t}\right)$ implies

$$E = e^{2N}, \quad \Phi(E) = \left(2Ne^2\right)^N, \quad \frac{\Phi(E)}{E} = (2N)^N, \quad \Phi'(E) = \frac{3}{2}(2N)^N.$$

Finally, we will use the estimate

$$\frac{1}{\widetilde{\Phi_N}^{(-1)}\left(\frac{1}{x}\right)} \leq \frac{(2\ln(2N))^N}{\left(\ln\frac{1}{x}\right)^N}, \qquad 0 < x \ll 1, \tag{4.5}$$

where $\widetilde{\Phi_N}^{(-1)}$ is the inverse of the conjugate Young function $\widetilde{\Phi_N}$. To see this, first note that we can write

$$\Phi(t) = \Phi_N(t) = t(\ln t)^N, \qquad t \geq e^{2N},$$

and therefore

$$\Phi'(t) = (\ln t)^N + tN(\ln t)^{N-1}\frac{1}{t} \geq (\ln t)^N, \qquad t \geq e^{2N}.$$

With $s = \Phi'(t)$, we then have

$$s \geq (\ln t)^N \text{ for } t \geq e^{2N}, \text{ i.e. } (\Phi')^{-1}(s) = t \leq e^{s^{\frac{1}{N}}} \text{ for } s \geq (2N)^N,$$

and thus

$$\begin{aligned} \widetilde{\Phi}'(s) &= (\Phi')^{-1}(s) \leq e^{s^{\frac{1}{N}}} \text{ for } s \geq (2N)^N, \\ \text{i.e. } \widetilde{\Phi}(s) &\leq Ns^{1-\frac{1}{N}}e^{s^{\frac{1}{N}}} \text{ for } s \geq (2N)^N, \end{aligned}$$

since

$$\begin{aligned} \frac{d}{ds}\left(Ns^{1-\frac{1}{N}}e^{s^{\frac{1}{N}}}\right) &= Ns^{1-\frac{1}{N}}e^{s^{\frac{1}{N}}}\frac{1}{N}s^{\frac{1}{N}-1} + N\left(1-\frac{1}{N}\right)s^{-\frac{1}{N}}e^{s^{\frac{1}{N}}} \\ &= e^{s^{\frac{1}{N}}} + (N-1)\,s^{-\frac{1}{N}}e^{s^{\frac{1}{N}}} \\ &= e^{s^{\frac{1}{N}}}\left\{1 + \frac{N-1}{s^{\frac{1}{N}}}\right\} \geq e^{s^{\frac{1}{N}}} \geq \frac{d}{ds}\widetilde{\Phi}(s). \end{aligned}$$

In order to estimate $\widetilde{\Phi}^{-1}(t)$, we write

$$t = \widetilde{\Phi}(s) \leq N s^{1-\frac{1}{N}} e^{s^{\frac{1}{N}}};$$

$$\ln t \leq \ln N + \left(1 - \frac{1}{N}\right) \ln s + s^{\frac{1}{N}} \leq (2\ln(2N)) s^{\frac{1}{N}}, \qquad \text{for all } \ s \geq (2N)^N;$$

$$\widetilde{\Phi}^{-1}(t) = s \geq \left(\frac{1}{2\ln(2N)} \ln t\right)^N, \qquad t \geq e^{2N}.$$

Then we obtain (4.5) by noting that

$$\frac{1}{\widetilde{\Phi}^{-1}(\frac{1}{x})} \leq \frac{1}{\left(\frac{1}{2\ln(2N)} \ln \frac{1}{x}\right)^N} = \frac{(2\ln(2N))^N}{\left(\ln \frac{1}{x}\right)^N}, \qquad 0 < x < e^{-2N}.$$

2. DeGiorgi iteration

In the next proposition we apply DeGiorgi iteration to a sequence of Orlicz Sobolev and Caccioppoli inequalities involving a family of bump functions adapted to the strongly degenerate geometries F_σ. Recall that the strongly degenerate geometries F_α have degeneracy function

$$f(x) = e^{-\frac{1}{x^\alpha}}, \quad \alpha > 0, \quad x \geq 0,$$

and that the family of bump functions $\{\Phi_N\}_{N>1}$ is given by

$$\Phi_N(t) \equiv \begin{cases} t(\ln t)^N & \text{if} \quad t \geq E = E_N = e^{2N} \\ (\ln E)^N t & \text{if} \quad 0 \leq t \leq E = E_N = e^{2N} \end{cases}. \tag{4.6}$$

PROPOSITION 27. *Assume that the Orlicz Sobolev norm inequality* (3.1) *holds with* $\Phi = \Phi_N$ *for some* $N > 1$ *and superradius* $\varphi(r)$, *and with a geometry* F *satisfying Definition* 16.

(1) *Then every weak subsolution* u *to* $\mathcal{L}u = \phi$ *in* Ω, *with* $\mathcal{L}$ *as in* (1.2), (1.3), *and* A*-admissible* ϕ, *satisfies the* inner ball inequality

$$\|u_+\|_{L^\infty(\frac{1}{2}B)} \leq A_{N,\varepsilon}(r) \left(\frac{1}{|B|}\int_B u_+^2\right)^{\frac{1}{2}} + \|\phi\|_X, \tag{4.7}$$

$$\text{where } A_{N,\varepsilon}(r) \equiv C_1 \exp\left\{C_2 \left(\frac{\varphi(r)}{r}\right)^{\frac{1}{N-1-\varepsilon}}\right\}, \tag{4.8}$$

for every ball $B \subset \Omega$ *with radius* r *that is centered on the* x_n*-axis, and every* $0 < \varepsilon < N - 1$. *Here the constants* C_1 *and* C_2 *depend on* N *and* ε *but not on* r.

(2) *Thus* u *is locally bounded above in* Ω, *since* $\mathcal{L}$ *is elliptic away from the* x_n*-axis by the structure conditions in Definition* 16.

(3) *In particular, weak solutions are locally bounded in* Ω.

PROOF. Without loss of generality, we may assume that $B = B(0, r)$ is a ball centered at the origin with radius $r > 0$. Let $\{\psi_j\}_{j=1}^\infty$ be a *standard* sequence of Lipschitz cutoff functions at $(0, r)$ as in Definition 7 with $\gamma = 1 + \frac{\varepsilon}{2}$, and associated with the balls $B_j \equiv B(0, r_j) \supset \text{supp}\psi_{j+1}$, $\psi_j = 1$ on B_j, with $r_1 = r$, $r_\infty \equiv \lim_{j\to\infty} r_j = \frac{r}{2}$, $r_j - r_{j+1} = \frac{c}{j^{1+\frac{\varepsilon}{2}}} r$ for a uniquely determined constant $c = c_\varepsilon$, and $\|\nabla_A \psi_j\|_\infty \lesssim \frac{j^{1+\frac{\varepsilon}{2}}}{r}$ with ∇_A as in (1.4) above (see Proposition 68 in [**SaWh4**] for more detail). Following DeGiorgi ([**DeG**], see also [**CaVa**]), we consider the family of truncations

$$u_k = (u - C_k)_+, \quad C_k = \tau \|\phi\|_X \left(1 - c(k+1)^{-\varepsilon/2}\right), \tag{4.9}$$

and denote the L^2 norm of the truncation u_k by

$$U_k \equiv \int_{B_k} |u_k|^2 d\mu, \tag{4.10}$$

where $d\mu = \frac{dx}{|B(0,r)|} = \frac{dx}{|B|}$ is independent of k. Here we have introduced a parameter $\tau \geq 1$ that will be used later for rescaling. We will assume $||\phi||_X > 0$, otherwise we replace it with parameter $m > 0$ and take the limit $m \to 0$ at the end of the argument.

Using Hölder's inequality for Young functions we can write

$$\int (\psi_{k+1} u_{k+1})^2 d\mu \leq C|| (\psi_{k+1} u_{k+1})^2 ||_{L^\Phi(B_k;\mu)} \cdot ||1||_{L^{\tilde{\Phi}}(\{\psi_{k+1} u_{k+1} > 0\};\mu)}, \tag{4.11}$$

where the norms are taken with respect to the measure μ. For the first factor on the right we have, using the Orlicz Sobolev inequality (3.1) and Cauchy-Schwartz inequality with δ to be determined later,

$$\begin{aligned}
&|| (\psi_{k+1} u_{k+1})^2 ||_{L^\Phi(B_k;\mu)} \\
&\quad \leq C\varphi(r) \int \left|\nabla_A \left(\psi_{k+1}^2 u_{k+1}^2\right)\right| d\mu = C\varphi(r) \int \left|\nabla_A \left(\psi_{k+1}^2 u_{k+1}^2\right)\right| d\mu \\
&\quad = C\varphi(r) \int |\psi_{k+1} u_{k+1}| \left|\nabla_A (\psi_{k+1} u_{k+1})\right| d\mu \\
&\quad \leq \delta \int \left|\nabla_A (\psi_{k+1} u_{k+1})\right|^2 d\mu + \frac{C}{\delta} \varphi(r)^2 \int \left(\psi_{k+1}^2 u_{k+1}^2\right) d\mu.
\end{aligned}$$

We would now like to apply the Caccioppoli inequality (3.8) with an appropriate function v to the first term on the right, and therefore we need to establish estimate (3.7). For that we observe that

$$\begin{aligned}
u_{k+1} &> 0 \Longrightarrow u > C_{k+1} = \tau \|\phi\|_X \left(1 - c(k+2)^{-\varepsilon/2}\right) &(4.12)\\
&\Longrightarrow u_k = (u - C_k)_+ > c\tau \|\phi\|_X \left[(k+1)^{-\frac{\varepsilon}{2}} - (k+2)^{-\frac{\varepsilon}{2}}\right], &(4.13)
\end{aligned}$$

and

$$\begin{aligned}
&c\tau \|\phi\|_X \left[(k+1)^{-\frac{\varepsilon}{2}} - (k+2)^{-\frac{\varepsilon}{2}}\right] = c\tau \|\phi\|_X (k+1)^{-\frac{\varepsilon}{2}} \left[1 - \left(\frac{k+1}{k+2}\right)^{\frac{\varepsilon}{2}}\right] \\
= \;& c\tau \|\phi\|_X (k+1)^{-\frac{\varepsilon}{2}} \left(1 - \frac{k+1}{k+2}\right) \frac{\varepsilon}{2} \theta^{\frac{\varepsilon}{2}-1} \quad \text{where } \frac{k+1}{k+2} < \theta < 1 \\
\geq \;& \frac{\varepsilon}{2} c\tau \|\phi\|_X (k+1)^{-\frac{\varepsilon}{2}} \frac{1}{k+2} \left(\frac{k+1}{k+2}\right)^{\frac{\varepsilon}{2}-1} \geq \frac{\varepsilon}{2} c\tau \|\phi\|_X (k+2)^{-1-\frac{\varepsilon}{2}}.
\end{aligned}$$

This implies

$$\|\phi\|_X \leq \frac{2}{c\tau\varepsilon} (k+2)^{1+\frac{\varepsilon}{2}} u_k \leq \frac{2}{c\varepsilon} (k+2)^{1+\frac{\varepsilon}{2}} u_k, \tag{4.14}$$

which is (3.7) with $P = \frac{2}{c\varepsilon} (k+2)^{1+\frac{\varepsilon}{2}}$ and $v = u_k$. Note that we have used our assumption that $\tau \geq 1$ in the display above.

Thus by the Caccioppoli inequality (3.8) we have,

$$\begin{aligned}\int \left|\nabla_A\left(\psi_{k+1}u_{k+1}\right)\right|^2 d\mu \\ &\leq C\left(\|\nabla_A\psi_{k+1}\|_{L^\infty}+\frac{2}{c\varepsilon}(k+2)^{1+\frac{\varepsilon}{2}}\right)^2\int_{B_k}\left(u_{k+1}^2+u_k^2\right)d\mu \\ &\leq C\frac{(k+1)^{2+\varepsilon}}{r^2}\int_{B_k}u_k^2 d\mu\ .\end{aligned}$$

Finally, choosing $\delta=\frac{r\varphi(r)}{(k+1)^{1+\frac{\varepsilon}{2}}}$, we obtain

(4.15)

$$\begin{aligned}\|\left(\psi_{k+1}u_{k+1}\right)^2\|_{L^\Phi(B_k;\mu)} &\leq \delta C\frac{(k+1)^{2+\varepsilon}}{r^2}\int_{B_k}u_k^2 d\mu+\frac{C}{\delta}\varphi(r)^2\int\left(\psi_{k+1}^2u_{k+1}^2\right)d\mu \\ &\leq C\frac{\varphi(r)}{r}(k+1)^{1+\frac{\varepsilon}{2}}\int_{B_k}u_k^2 d\mu\ .\end{aligned}$$

For the second factor in (4.11) we claim

$$\|1\|_{L^{\tilde\Phi}(\{\psi_{k+1}u_{k+1}>0\}d\mu)}\leq\Gamma\left(\left|\left\{\psi_k u_k>\frac{\varepsilon}{2}c(k+2)^{-1-\frac{\varepsilon}{2}}\right\}\right|_\mu\right), \tag{4.16}$$

with the notation

$$\Gamma(t)\equiv\frac{1}{\tilde\Phi^{-1}\left(\frac{1}{t}\right)}. \tag{4.17}$$

First recall

$$\|f\|_{L^{\tilde\Phi}(X)}\equiv\inf\left\{a:\int_X\tilde\Phi\left(\frac{f}{a}\right)\leq 1\right\},$$

and note

$$\int_{\{\psi_{k+1}u_{k+1}>0\}}\tilde\Phi\left(\frac{1}{a}\right)=\tilde\Phi\left(\frac{1}{a}\right)\left|\{\psi_{k+1}u_{k+1}>0\}\right|_\mu .$$

Now take

$$a=\Gamma\left(\left|\{\psi_{k+1}u_{k+1}>0\}\right|_\mu\right)\equiv\frac{1}{\tilde\Phi^{-1}\left(\frac{1}{\left|\{\psi_{k+1}u_{k+1}>0\}\right|_\mu}\right)}$$

which obviously satisfies

$$\int_{\{\psi_{k+1}u_{k+1}>0\}}\tilde\Phi\left(\frac{1}{a}\right)d\mu=1.$$

This gives

$$\|1\|_{L^{\tilde\Phi}(\{\psi_{k+1}u_{k+1}>0\}d\mu)}\leq a=\Gamma\left(\left|\{\psi_{k+1}u_{k+1}>0\}\right|_\mu\right),$$

and to conclude (4.16) we only need to observe that

$$\{\psi_{k+1}u_{k+1}>0\}\subset\left\{\psi_{k+1}u_k>\frac{\varepsilon}{2}c\tau\|\phi\|_X(k+2)^{-1-\frac{\varepsilon}{2}}\right\},$$

which follows from (4.12).

Next we use Chebyshev's inequality to obtain

$$\left|\left\{\psi_{k+1}u_k>\frac{\varepsilon}{2}c(k+2)^{-1-\frac{\varepsilon}{2}}\right\}\right|_\mu\leq\gamma(k+2)^{2+\varepsilon}\int\left(\psi_k u_k\right)^2 d\mu, \tag{4.18}$$

where $\gamma = \frac{4}{c^2\tau^2\|\phi\|_X^2\varepsilon^2}$. Combining (4.11)-(4.18) we obtain

$$\begin{aligned}
&\int (\psi_{k+1}u_{k+1})^2\,d\mu \\
&\leq C\|\,(\psi_{k+1}u_{k+1})^2\,\|_{L^{\Phi}(B_k;\mu)}\cdot\|1\|_{L^{\tilde{\Phi}}(\{\psi_{k+1}u_{k+1}>0\};\mu)} \\
&\leq C\frac{\varphi(r)}{r}\frac{(k+1)^{1+\frac{\varepsilon}{2}}}{|B_1|}\int_{B_k}u_k^2\cdot\Gamma\left(\left|\left\{\psi_k u_k>\frac{\varepsilon}{2}c\,(k+2)^{-1-\frac{\varepsilon}{2}}\right\}\right|_\mu\right) \\
&\leq C\frac{\varphi(r)}{r}(k+1)^{1+\frac{\varepsilon}{2}}\left(\int_{B_k}u_k^2 d\mu\right)\Gamma\left(\gamma\,(k+2)^{2+\varepsilon}\int(\psi_k u_k)^2\,d\mu\right) \\
&= C\frac{\varphi(r)}{r}(k+1)^{1+\frac{\varepsilon}{2}}\left(\int_{B_k}u_k^2 d\mu\right)\Gamma\left(\gamma\,(k+2)^{2+\varepsilon}\int(\psi_k u_k)^2\,d\mu\right),
\end{aligned}$$

or in terms of the quantities U_k,

$$\begin{aligned}
U_{k+1} &\leq C\frac{\varphi(r)}{r}(k+1)^{1+\frac{\varepsilon}{2}}U_k\Gamma\left(\gamma\,(k+2)^{2+\varepsilon}U_k\right) \\
&= C\frac{\varphi(r)}{r}(k+1)^{1+\frac{\varepsilon}{2}}U_k\frac{1}{\tilde{\Phi}^{-1}\left(\frac{1}{\gamma(k+2)^{2+\varepsilon}U_k}\right)}.
\end{aligned}\tag{4.19}$$

Now we use the estimate (4.5) on $\frac{1}{\tilde{\Phi}^{-1}\left(\frac{1}{x}\right)}$ to determine the values of N and ε for which DeGiorgi iteration provides local boundedness of weak subsolutions, i.e. for which $U_k\to 0$ as $k\to\infty$ provided U_0 is small enough. From (4.5) and (4.19) we have

$$U_{k+1}\leq C\frac{\varphi(r)}{r}\frac{(k+1)^{1+\frac{\varepsilon}{2}}U_k}{\left(\ln\frac{1}{\gamma}\frac{1}{(k+2)^{(2+\varepsilon)}U_k}\right)^N},$$

provided

$$\gamma\,(k+2)^{(2+\varepsilon)}U_k<e^{-2N},$$

and using the notation

$$b_k\equiv\ln\frac{1}{U_k},$$

we can rewrite this as

$$\begin{aligned}
b_{k+1} &\geq b_k-(1+\frac{\varepsilon}{2})\ln(k+1) \\
&\quad +N\ln(b_k-(2+\varepsilon)\ln(k+2)-\ln\gamma)-\ln\left(C\frac{\varphi(r)}{r}\right),
\end{aligned}$$

for $k\geq 0$, provided

$$\frac{1}{\gamma}(k+2)^{-(2+\varepsilon)}\frac{1}{U_k}>e^{2N};$$

i.e.

$$b_k>(2+\varepsilon)\ln(k+2)+\ln\gamma+2N,\qquad k\geq 0.\tag{4.20}$$

We now use induction to show that

Claim: Both (4.20) and

$$b_k\geq b_0+k,\qquad k\geq 0,\tag{4.21}$$

hold for b_0 taken sufficiently large depending on $N>1$ and $0<r<R$.

Indeed, both (4.20) and (4.21) are trivial if $k = 0$ and b_0 is large enough. Assume now that the claim is true for some $k \geq 0$. Then

$$\begin{aligned} b_{k+1} &\geq b_0 + k + 1 + N \ln (b_0 + k - (2+\varepsilon) \ln (k+2) - \ln \gamma) \\ &\quad -(1+\frac{\varepsilon}{2}) \ln (k+1) - 1 - \ln \left(C \frac{\varphi(r)}{r} \right). \end{aligned}$$

Now for $N > 1 + \frac{\varepsilon}{2}$ we have

$$N \ln (b_0 + k - (2+\varepsilon) \ln (k+2) - \ln \gamma) - (1+\frac{\varepsilon}{2}) \ln (k+1) - 1 - \ln \left(C \frac{\varphi(r)}{r} \right) \to \infty,$$

as $k \to \infty$, and therefore for b_0 sufficiently large depending on N and r, we obtain

$$N \ln (b_0 + k - (2+\varepsilon) \ln (k+2) - \ln \gamma) - (1+\frac{\varepsilon}{2}) \ln (k+1) - 1 - \ln \left(C \frac{\varphi(r)}{r} \right) \geq 0,$$

for all $k \geq 1$, which gives (4.21) for $k+1$,

$$b_{k+1} \geq b_0 + k + 1,$$

and also (4.20) for $k+1$,

$$\begin{aligned} b_{k+1} &\geq b_0 + k + 1 > (2+\varepsilon) \ln 2 + \ln \gamma + 2N + k + 1 \\ &\geq (2+\varepsilon) \ln (k+3) + \ln \gamma + 2N. \end{aligned}$$

We note that it is sufficient to require

$$(N - 1 - \varepsilon) \ln (b_0 - 8 - \ln \gamma) \geq \ln \left(C \frac{\varphi(r)}{r} \right),$$

or

$$b_0 \geq A_{N,\varepsilon} \left(\frac{\varphi(r)}{r} \right)^{\frac{1}{N-1-\varepsilon}} + \ln(e^8 \gamma) = A_{N,\varepsilon} \left(\frac{\varphi(r)}{r} \right)^{\frac{1}{N-1-\varepsilon}} + \ln \frac{4e^8}{c^2 \tau^2 \|\phi\|_X^2 \varepsilon^2} \tag{4.22}$$

for $A_{N,\varepsilon}$ sufficiently large depending on N and ε. This completes the proof of the induction step and therefore $b_k \to \infty$ as $k \to \infty$, or $U_k \to 0$ as $k \to \infty$, provided $U_0 = e^{-b_0}$ is sufficiently small.

Altogether, we have shown that

$$u_\infty = (u - \tau \|\phi\|_X)_+ = 0 \quad \text{on } B_\infty = B\left(0, \frac{r}{2}\right) = \frac{1}{2} B(0,r),$$

and thus that

$$u \leq \tau \|\phi\|_X \quad \text{on } B_\infty, \tag{4.23}$$

provided $U_0 = \int_B |u_0|^2 d\mu = \frac{1}{|B|} \int_B |u_0|^2$ is sufficiently small. From (4.22) it follows that it is sufficient to require

$$\sqrt{\frac{1}{|B|} \int_B |u_0|^2} = e^{-b_0} \leq \tau \|\phi\|_X C \varepsilon \exp \left(-\frac{A_{N,\varepsilon}}{2} \left(\frac{\varphi(r)}{r} \right)^{\frac{1}{N-1-\varepsilon}} \right) = \eta_{N,\varepsilon}(r) \tau \|\phi\|_X, \tag{4.24}$$

where $A_{N,\varepsilon}$ is the constant in (4.22), $C = c/2e^4$ and

$$\eta_{N,\varepsilon}(r) \equiv C \varepsilon \exp \left\{ -\frac{A_{N,\varepsilon}}{2} \left(\frac{\varphi(r)}{r} \right)^{\frac{1}{N-1-\varepsilon}} \right\}.$$

To recover the general case we now consider two cases, $\sqrt{\frac{1}{|B|}\int_B |u_0|^2} \leq \eta_{N,\varepsilon}(r) \|\phi\|_X$ and $\sqrt{\frac{1}{|B|}\int_B |u_0|^2} > \eta_{N,\varepsilon}(r) \|\phi\|_X$. In the first case we obtain that (4.24) holds with $\tau = 1$ and thus

$$\|u_+\|_{L^\infty(\frac{1}{2}B)} \leq \|\phi\|_X \ .$$

In the second case, when $\sqrt{\frac{1}{|B|}\int_B |u_0|^2} > \eta_{N,\varepsilon}(r) \|\phi\|_X$, we let $\tau = \frac{\sqrt{\frac{1}{|B|}\int_B |u_0|^2}}{\eta_{N,\varepsilon}(r)\|\phi\|_X} > 1$ so that (4.24) holds, and then from (4.23) we get

$$\|u_+\|_{L^\infty(\frac{1}{2}B)} \leq \frac{\sqrt{\frac{1}{|B|}\int_B |u_0|^2}}{\eta_{N,\varepsilon}(r)}.$$

Altogether this gives

$$\|u_+\|_{L^\infty(\frac{1}{2}B)} \leq \|\phi\|_X + A_{N,\varepsilon}(r)\sqrt{\frac{1}{|B|}\int_B |u_0|^2},$$

where

$$A_{N,\varepsilon}(r) = \frac{1}{\eta_{N,\varepsilon}(r)} = C_1 \exp\left(C_2 \left(\frac{\varphi(r)}{r} \right)^{\frac{1}{N-1-\varepsilon}} \right),$$

and where the constants C_1 and C_2 depend on N and ε, but do not depend on r. □

The following corollary makes the somewhat trivial observation that we may replace the factor $\frac{1}{|B|}$ with the much smaller factor $\frac{1}{|3B|}$ at the expense of replacing the constant $A_{N,\varepsilon}(r)$ with the much larger constant $A_{N,\varepsilon}(3r)$. This turns out to be a convenient renormalization of the local boundedness inequality for use in our continuity theorem below.

COROLLARY 28 (of the proof). *Suppose all the assumptions of Proposition 27 are satisfied. Then*

$$\|u_+\|_{L^\infty(\frac{1}{2}B)} \leq A_{N,\varepsilon}(3r) \left(\frac{1}{|3B|} \int_B u_+^2 \right)^{\frac{1}{2}} + \|\phi\|_X \ , \tag{4.25}$$

with $A_{N,\varepsilon}$ as in (4.8).

PROOF. The standard cutoff functions $\{\psi_k\}$ defined in the proof of Proposition 4.7 can also be considered as cutoff functions supported on $3B = B(0,3r)$. Thus we can repeat the proof of the proposition but applying the Orlicz Sobolev inequality relative to $3B$ instead of B. This results in changing the measure $d\mu = \frac{dx}{|B|}$ to the measure $d\mu = \frac{dx}{|3B|}$, and in changing the superradius ratio $\frac{\varphi(r)}{r}$ to $\frac{\varphi(3r)}{3r}$. The remaining estimates are unchanged since the cutoff functions are still only supported inside B, so the regions of integration remain unchanged. □

CHAPTER 5

Maximum Principle

Now we turn to the abstract maximum principle for weak subsolutions. We will assume that $f(x) \neq 0$ if $x \neq 0$, and that F satisfies the five geometric structure conditions in Definition 16. We also assume the following global Orlicz Sobolev inequality

$$\|w\|_{L^{\Phi}(\Omega)} \leq C(\Omega) \|\nabla_A w\|_{L^1(\Omega)}, \qquad w \in \left(W_A^{1,1}\right)_0(\Omega) \tag{5.1}$$

where $\Phi = \Phi_N$ with $N > 1$ as defined in (4.6).

The second ingredient of the proof is the following Caccioppoli inequality, which we show follows from the proof of Proposition 23 similar to Corollary 25.

PROPOSITION 29. *Let $u \in \left(W_A^{1,2}\right)_0(B)$ be a weak subsolution to $\mathcal{L}u = \phi$ with $\mathcal{L}$ as in (1.2), (1.3), and A-admissible ϕ, and suppose for some ball B, a constant $P > 0$, and a nonnegative function $v \in W_A^{1,2}(B)$ there holds*

$$\|\phi\|_X^2 \leq Pv(x), \quad a.e.\ x \in \{u > 0\} \cap B. \tag{5.2}$$

Then

$$\int_B |\nabla_A u_+|^2 d\mu \leq CP^2 \int_B v^2 d\mu, \tag{5.3}$$

where $d\mu = \frac{dx}{|B|}$, and $C = 1/c$ with c from (1.3). The same inequality holds with u and u_- in place of u_+.

PROOF. Since $w \equiv u_+ \in \left(W_A^{1,2}\right)_0(B)$ and u is a weak subsolution to $\mathcal{L}u = \phi$ we have

$$\begin{aligned}\int \nabla(u_+) \mathcal{A}\nabla u_+ d\mu &= \int \nabla u_+ \mathcal{A}\nabla u d\mu \leq -\int u_+ \phi d\mu \\ &\leq \|\phi\|_X \int |\nabla_{\mathcal{A}} u_+| \, d\mu \leq P \int v \, |\nabla_{\mathcal{A}} u_+| \, d\mu \,,\end{aligned}$$

where for the last inequality we used condition (5.2). Using Hölder's inequality and (1.3) this gives

$$\int |\nabla_A u_+|^2 \, d\mu \leq \frac{1}{c} \int |\nabla_{\mathcal{A}} u_+|^2 \, d\mu \leq CP^2 \int v^2 d\mu,$$

which is (5.3). Using $w = u$ or $w = u_-$ as a test function we obtain the last statement of the Proposition. □

We are now ready to prove the maximum principle.

THEOREM 30. *Let Ω be a bounded open subset of $\mathbb{R}^n$, and assume the global Orlicz Sobolev inequality (5.1) holds for some $N > 1$. Assume that u is a weak subsolution to $\mathcal{L}u = \phi$ in Ω with A-admissible ϕ, and that u is bounded on the boundary $\partial\Omega$. Then the following maximum principle holds,*

$$\sup_{\Omega} u \leq \sup_{\partial\Omega} u + C \left\|\phi\right\|_{X(\Omega)} ,$$

and in particular u is globally bounded.

PROOF. First, we proceed similarly to the proof of Proposition 27. Define

$$u_k \equiv (u - \tau \sup_{\partial\Omega} u - C_k)_+, \quad C_k = \tau \left\|\phi\right\|_X \left(1 - c\,(k+1)^{-\varepsilon/2}\right), \quad \tau \geq 1,$$

and note that $u_k \equiv 0$ for all k on $\partial\Omega$, so we can formally take $\psi_k \equiv 1$ for all k. Let $B = B(0, r_0)$ be a ball containing Ω and extend u_k to be zero outside Ω. Then we can assume $u_k \in \left(W_A^{1,2}\right)_0(B)$ and therefore Caccioppoli inequality (5.3) holds. Using (5.3) and the bound (4.14) on $\{x \in \Omega : u_{k+1} > 0\}$ we obtain the following version of inequality (4.19)

$$U_{k+1} \leq C(k+2)^{1+\varepsilon/2} U_k \frac{1}{\tilde{\Phi}^{-1}\left(\frac{1}{\gamma(k+2)^{2+\varepsilon} U_k}\right)},$$

where $U_k = \int_\Omega |u_k|^2$. Proceeding exactly as in the proof of Proposition 27 we thus obtain

$$u_\infty = (u - \tau \sup_{\partial\Omega} u - \tau \left\|\phi\right\|_X)_+ = 0,$$

provided U_0 is sufficiently small, and by the same argument as in the proof of Proposition 27

$$||u||_{L^\infty(\Omega)} \leq ||\phi||_X + \sup_{\partial\Omega} u + C||u||_{L^2(\Omega)}.$$

Finally, we use Caccioppoli inequality together with Orlicz Sobolev inequality to bound the last term

$$||u||_{L^2}^2 \leq ||u^2||_{L^\Phi} \leq C||\nabla_A u||_{L^2}^2 \leq C||\phi||_X,$$

where we used (5.3) with u in place of u_+, $v \equiv ||\varphi||_X$, and $P = 1$. □

CHAPTER 6

Continuity

Here we turn to proving the abstract continuity Theorem 21. The key tool for this is an infinitely degenerate variation on the DeGiorgi Lemma in Lemma 1.4 of Caffarelli and Vasseur [**CaVa**], but yielding an estimate different from that of Caffarelli and Vasseur - one that does not involve an isoperimetric inequality. For convenience we recall the proportional vanishing L^1-Sobolev inequality (2.2) that we are assuming in the abstract continuity Theorem 21:

$$\int_B |w| \leq Cr(B) \int_{2B} |\nabla_A w|, \tag{6.1}$$

for all Lipschitz w supported in $2B$ that vanish on a subset E of a ball B with $|E| \geq \frac{1}{2}|B|$.

LEMMA 31. *Suppose that the proportional vanishing L^1-Sobolev inequality (6.1) holds. Fix x and r and suppose that w satisfies $\int_{B(x,2r)} |\nabla_A w_+(y)|^2 dy \leq C_0$. Set*

$$\begin{aligned}
\mathcal{A} &\equiv \{y \in B(x,r) : w(y) \leq 0\}, \\
\mathcal{C} &\equiv \{z \in B(x,r) : w(z) \geq 1\}, \\
\mathcal{D} &\equiv \{y \in B(x,2r) : 0 < w(y) < 1\}.
\end{aligned}$$

Then if $|\mathcal{A}| \geq \frac{1}{2}|B(x,r)|$, we have

$$C_0 |\mathcal{D}| \geq C_1 \left(\frac{|\mathcal{A}| |\mathcal{C}|}{r |B(x,r)|} \right)^2. \tag{6.2}$$

PROOF. Let $\overline{w}(y) \equiv \max\{0, \min\{1, w(y)\}\}$, and note that $\overline{w}(z) = 1$ for $z \in \mathcal{C}$. Then applying (6.1) with $w = \overline{w}$, $B = B(x, r_0)$ and $E = \mathcal{A}$, we have that

$$\begin{aligned}
|\mathcal{C}| |\mathcal{A}| &= \int_{\mathcal{C}} \overline{w}(z)\, dz \ |\mathcal{A}| \leq \int_{B(x,r)} \overline{w}(z)\, dz \ |\mathcal{A}| \\
&\leq Cr |B(x,r)| \int_{B(x,2r)} |\nabla_A \overline{w}(y)|\, dy = Cr |B(x,r)| \int_{\mathcal{D}} |\nabla_A w(y)|\, dy \\
&\lesssim r_0 |B(x,r)| \sqrt{|\mathcal{D}|}\, \|\nabla_A w\|_{L^2(B(x,2r))} \lesssim \sqrt{C_0} r |B(x,r)| \sqrt{|\mathcal{D}|}.
\end{aligned}$$

Thus we obtain

$$|\mathcal{C}| |\mathcal{A}| \lesssim r |B(x,r)| \sqrt{C_0 |\mathcal{D}|},$$

or

$$C_0 |\mathcal{D}| \geq C_1 \left(\frac{|\mathcal{A}| |\mathcal{C}|}{r |B(x,r)|} \right)^2.$$

□

We can now prove Theorem 21 for weak solutions to a homogeneous degenerate equation. In fact it is easily seen, using Corollary 28, that it suffices to prove the following local statement with $\frac{1}{2\sqrt{\delta(r)}} = A_{N,\varepsilon}(3r)$, where $A_{N,\varepsilon}(r)$ is the constant in the local boundedness inequalities (4.7) and (4.25).

PROPOSITION 32. *Let $B_r = B(x,r)$. Suppose that (6.2) holds, and also that for some $\delta(r) > 0$, we have the following local boundedness inequality,*

$$(6.3)\quad \|u_+\|_{L^\infty\left(B_{\frac{r}{2}}\right)} \leq \frac{1}{2\sqrt{\delta(r)}}\left(\frac{1}{|B_{3r}|}\int_{B_r} u_+^2\right)^{\frac{1}{2}}, \qquad \textit{whenever } Lu = 0 \textit{ in } B_r\ ,$$

for all $0 < r < r_0$. Moreover we assume the summability condition

$$(6.4)\qquad \sum_{j=1}^{\infty} \lambda_j = \infty,$$

where $r_j = \frac{r_0}{4^j}$ and $\lambda_j \equiv \frac{1}{2^{3+\frac{C_3}{\delta^2(r_j)}}}$ for $j \geq 1$, and where C_3 is a positive constant. Then if u is a weak solution to $\mathcal{L}u = 0$ in B_{r_0}, we conclude that u is continuous at x.

We now reduce the proof of Proposition 32 to proving Proposition 34 below, which shows that if a solution v is bounded by 1 in a ball B_{3r}, and is nonpositive on a 'sufficiently large' subset of the smaller ball B_r, then v is in fact bounded by a constant *less* than 1 in the smaller ball B_r. This reduction is achieved in two steps by way of the following lemma.

LEMMA 33. *Suppose that the local boundedness property in Proposition 32 holds. Let u be a weak solution of $\mathcal{L}u = 0$ in B_{r_0}. Then with $\lambda(r) \equiv \frac{1}{2^{3+\frac{C_3}{\delta(r)^2}}}$ we have*

$$\operatorname{osc}_{B_{\frac{r}{2}}} u \leq \left(1 - \frac{\lambda(r)}{2}\right)\operatorname{osc}_{B_r} u, \qquad 0 < r < r_0.$$

Indeed, iterating Lemma 33 gives

$$\operatorname{osc}_{B_{\frac{r}{4^\ell}}} u \leq \left\{\prod_{j=1}^{\ell}\left(1 - \frac{\lambda\left(\frac{r}{4^j}\right)}{2}\right)\right\}\operatorname{osc}_{B_r} u, \qquad \ell \geq 1,$$

which implies $\lim_{\ell\to\infty} \operatorname{osc}_{B_{\frac{r}{4^\ell}}} u = 0$ since the infinite product above vanishes if the summability condition (6.4) holds. Hence u is continuous at x, which proves Proposition 32.

Next we note that Lemma 33 is in turn an easy consequence of this proposition.

PROPOSITION 34. *Suppose that the local boundedness property (2.5) holds. With r sufficiently small, let*

$$\lambda(r) \equiv \frac{1}{2^{3+\frac{C_3}{\delta(r)^2}}} \in (0,1)\,.$$

Let $v \leq 1$ and $\mathcal{L}v = 0$ in B_{3r}. Assume that $|B_r \cap \{v \leq 0\}| \geq \frac{1}{2}|B_r|$. Then

$$\sup_{B_{\frac{r}{2}}} v \leq 1 - \lambda(r)\,.$$

Indeed, to prove Lemma 33 from Proposition 34, let

$$v(x) \equiv \frac{2}{\operatorname{osc}_{B_r} u} \left\{ u(x) - \frac{\sup_{B_r} u + \inf_{B_r} u}{2} \right\},$$

so that $-1 \leq v \leq 1$ on B_r. If $|\{v \leq 0\} \cap B_r| \geq \frac{1}{2}|B_r|$, then Proposition 34 gives $\operatorname{osc}_{B_{\frac{r}{2}}} v \leq 2 - \lambda(r)$, which implies

$$\operatorname{osc}_{B_{\frac{r}{2}}} u \leq \left(1 - \frac{\lambda(r)}{2}\right) \operatorname{osc}_{B_r} u.$$

If instead we have $|\{v \geq 0\} \cap B_r| \geq \frac{1}{2}|B_r|$, we obtain the same inequality by applying the above argument to $-v$. This completes the proof of Lemma 33.

Thus matters have been reduced to proving Proposition 34.

PROOF OF PROPOSITION 34. Define

$$w_k = 2^k \left(v - (1 - 2^{-k})\right)$$

and let $\varphi \in C_0^\infty(B_{3r})$ be such that $\varphi = 1$ on B_{2r} and $||\nabla_A \varphi||_{L^\infty(B_{3r})} \leq \frac{2}{r}$. Using $\mathcal{L}w_k = 0$ in B_{3r_0} and the standard Caccioppoli inequality (3.2) with $\phi = 0$ we have

$$\begin{aligned}\int_{B_{2r}} |\nabla_A (w_k)_+|^2 \leq \int_{B_{3r}} |\nabla_A (\varphi (w_k)_+)|^2 &\leq C ||\nabla_A \varphi||_{L^\infty}^2 \int_{B_{3r} \cap supp \varphi} (w_k)_+^2 \\ &\leq 4C\,|B_{3r}|\, r^{-2},\end{aligned}$$

where the last inequality follows from the fact that $w_k \leq 1$, which in turn follows from the assumption that $v \leq 1$. Define

$$\begin{aligned}\mathcal{A}_k &= \{2w_k \leq 0\} \cap B_r, \\ \mathcal{C}_k &= \{2w_k \geq 1\} \cap B_r, \\ \mathcal{D}_k &= \{0 < 2w_k < 1\} \cap B_{2r}.\end{aligned}$$

Also note that $v \leq 0$ implies $w_k \leq 0$, and therefore we have

$$|\mathcal{A}_k| = |\{2w_k \leq 0\} \cap B_r| = |\{w_k \leq 0\} \cap B_r| \geq \frac{1}{2}|B_r|, \tag{6.5}$$

so that the hypothesis $|\mathcal{A}_k| \geq \frac{1}{2}|B_r|$ in the proportional vanishing L^1-Sobolev inequality (6.1) holds.

We will apply Lemma 31 with $w = 2w_k$ recursively for $k = 0, 1, 2, \ldots$ below, but only as long as

$$\int_{B_r} (w_{k+1})_+^2 \frac{dx}{|B_{3r}|} \geq \delta(r), \tag{6.6}$$

where $\delta(r)$ is the positive constant in Proposition 32. We use both (6.5) and (6.6), and the fact that

$$w_{k+1} = 2w_k - 1$$

implies $\mathcal{C}_k = \{w_{k+1} \geq 0\} \cap B_r$, and hence

$$|\mathcal{C}_k| \geq \int_{\mathcal{C}_k} (w_{k+1})_+^2 dx = \int_{B_r} (w_{k+1})_+^2 dx \geq \delta(r)\,|B_{3r}|,$$

to obtain from Lemma 31 that

$$\begin{aligned}\left|\left\{0<w_k<\frac{1}{2}\right\}\cap B_{2r}\right| &= \left|\{0<2w_k<1\}\cap B_{2r}\right| = |\mathcal{D}_k| \\ &\geq \frac{C_1}{C\,|B_{3r}|\,r^{-2}}\left(\frac{\frac{1}{2}\,|B_r|\,\delta(r)\,|B_{3r}|}{r\,|B_r|}\right)^2 \\ &= \frac{C_1}{4C}\delta(r)^2\,|B_{3r}| = \alpha(r),\end{aligned}$$

where

$$\alpha(r) \equiv \frac{C_1}{4C}\delta(r)^2\,|B_{3r}| > 0 \tag{6.7}$$

depends on r, but not on k or v. This gives

$$\begin{aligned}|B_{2r}| &\geq |\{w_k \leq 0\}\cap B_{2r}| = |\{2w_{k-1}\leq 1\}\cap B_{2r}| \\ &= |\{w_{k-1}\leq 0\}\cap B_{2r}| + \left|\left\{0<w_{k-1}\leq\frac{1}{2}\right\}\cap B_{2r}\right| \\ &\geq |\{w_{k-1}\leq 0\}\cap B_{2r}| + \alpha \\ &\;\;\vdots \\ &\geq |\{w_0\leq 0\}\cap B_{2r}| + k\alpha(r) \geq k\alpha(r),\end{aligned}$$

The above inequality, namely $|B_{2r}| \geq k\alpha(r)$, must fail for a finite k *independent* of v, in fact it fails for the unique integer $k_0 \geq 0$ satisfying $\frac{|B_{2r}|}{\alpha(r)} < k_0 \leq \frac{|B_{2r}|}{\alpha(r)} + 1$, and for this k_0 we must then have

$$\int_{B_r} (w_{k_0+1})_+^2\,dx < \delta(r)\,|B_{3r}|.$$

By the local boundedness inequality (2.5), we conclude that in the ball $B_{\frac{r}{2}}$, we have

$$\begin{aligned}w_{k_0+1} \leq \left\|(w_{k_0+1})_+\right\|_{L^\infty\left(B_{\frac{r}{2}}\right)} &\leq \frac{1}{2\sqrt{\delta(r)}}\left(\frac{1}{|B_{3r}|}\int_{B_r}(w_{k_0+1})_+^2\right)^{\frac{1}{2}} \\ &< \frac{1}{2\sqrt{\delta(r)}}\sqrt{\delta(r)} = \frac{1}{2}.\end{aligned}$$

Rescaling back to v now gives

$$\begin{aligned} 2^{k_0+1}\left[v-\left(1-\frac{1}{2^{k_0+1}}\right)\right]_+ &= (w_{k_0+1})_+ \leq \frac{1}{2} \text{ in } B_{\frac{r}{2}} \\ \Longrightarrow \qquad v \leq 1-\frac{1}{2^{k_0+1}}+\frac{1}{2}\frac{1}{2^{k_0+1}} &= 1-\frac{1}{2^{k_0+2}} \text{ in } B_{\frac{r}{2}}.\end{aligned}$$

Finally, we note that (6.7) gives $\alpha(r) = \frac{C_1}{4C}\delta(r)^2\,|B_{3r}|$, and hence

$$k_0 \leq \frac{|B_{2r}|}{\alpha(r)} + 1 = \frac{|B_{2r}|}{\frac{C_1}{4C}\delta(r)^2\,|B_{3r}|} + 1 \leq \frac{C_3}{\delta(r)^2} + 1,$$

and thus that

$$\frac{1}{2^{k_0+2}} \geq \frac{1}{2^{3+\frac{C_3}{\delta(r)^2}}} = \lambda(r).$$

□

Part 3

Geometric theory

In this third part of the paper, we turn to the problem of finding specific geometric conditions on the structure of our equations that permit us to prove the Orlicz Sobolev and proportional vanishing L^1 inequalities needed to apply the abstract theory in Part 2 above. The first chapter here deals with basic geometric estimates for a specific family of geometries, which are then applied in the next chapter to obtain the needed Orlicz Sobolev and proportional vanishing L^1 inequalities. Finally, in the third chapter in this part we prove our geometric theorems on local boundedness, the maximum principle and continuity of weak solutions.

CHAPTER 7

Infinitely Degenerate Geometries

In this first chapter of the third part of the paper, we begin with degenerate geometries in the plane, the properties of their geodesics and balls, and the associated subrepresentation inequalities. The final chapter will treat higher dimensional geometries. Recall from (2.4) that we are considering the inverse metric tensor given by the 2×2 diagonal matrix

$$A = \begin{bmatrix} 1 & 0 \\ 0 & f(x)^2 \end{bmatrix}.$$

Here the function $f(x)$ is an even twice continuously differentiable function on the real line $\mathbb{R}$ with $f(0) = 0$ and $f'(x) > 0$ for all $x > 0$. The A-distance dt is given by

$$dt^2 = dx^2 + \frac{1}{f(x)^2} dy^2.$$

This distance coincides with the control distance as in [**SaWh4**], etc. since a vector v is subunit for an invertible symmetric matrix A, i.e. $(v \cdot \xi)^2 \leq \xi^{\mathrm{tr}} A \xi$ for all ξ, if and only if $v^{\mathrm{tr}} A^{-1} v \leq 1$. Indeed, if v is subunit for A, then

$$\left(v^{\mathrm{tr}} A^{-1} v\right)^2 = \left(v \cdot A^{-1} v\right)^2 \leq \left(A^{-1} v\right)^{\mathrm{tr}} A A^{-1} v = v^{\mathrm{tr}} A^{-1} v,$$

and for the converse, Cauchy-Schwartz gives

$$(v \cdot \xi)^2 = \left(v^{\mathrm{tr}} \xi\right)^2 = \left(v^{\mathrm{tr}} A^{-1} A \xi\right)^2 \leq \left(v^{\mathrm{tr}} A^{-1} A A^{-1} v\right) \left(\xi^{\mathrm{tr}} A \xi\right) = \left(v^{\mathrm{tr}} A^{-1} v\right) \left(\xi^{\mathrm{tr}} A \xi\right).$$

1. Calculation of the A-geodesics

We now compute the equation satisfied by an A-geodesic γ passing through the origin. A geodesic minimizes the distance

$$\int_0^{x_0} \sqrt{1 + \frac{\left(\frac{dy}{dx}\right)^2}{f^2}}\, dx, \qquad \text{where } (x, y) \text{ is on } \gamma,$$

and so the calculus of variations gives the equation

$$\frac{d}{dx} \left[\frac{\frac{dy}{dx}}{f^2 \sqrt{1 + \frac{\left(\frac{dy}{dx}\right)^2}{f^2}}} \right] = 0.$$

Consequently, the function

$$\lambda = \frac{f^2 \sqrt{1 + \frac{\left(\frac{dy}{dx}\right)^2}{f^2}}}{\frac{dy}{dx}}$$

is actually a positive constant conserved along the geodesic $y = y(x)$ that satisfies

$$\lambda^2 = \frac{f^2\left[f^2 + \left(\frac{dy}{dx}\right)^2\right]}{\left(\frac{dy}{dx}\right)^2} \Longrightarrow \left(\lambda^2 - f^2\right)\left(\frac{dy}{dx}\right)^2 = f^4.$$

Thus if $\gamma_{0,\lambda}$ denotes the geodesic starting at the origin going in the vertical direction for $x > 0$, and parameterized by the constant λ, we have $\lambda = f(x)$ if and only if $\frac{dy}{dx} = \infty$. For convenience we temporarily assume that f is defined on $(0, \infty)$. Thus the geodesic $\gamma_{0,\lambda}$ turns back toward the y-axis at the unique point $(X(\lambda), Y(\lambda))$ on the geodesic where $\lambda = f(X(\lambda))$, provided of course that $\lambda < f(\infty) \equiv \sup_{x>0} f(x)$. On the other hand, if $\lambda > f(\infty)$, then $\frac{dy}{dx}$ is essentially constant for x large and the geodesics $\gamma_{0,\lambda}$ for $\lambda > f(\infty)$ look like straight lines with slope $\frac{f(\infty)^2}{\sqrt{\lambda^2 - f(\infty)^2}}$ for x large. Finally, if $\lambda = f(\infty)$, then the geodesic $\gamma_{0,\lambda}$ has slope that blows up at infinity.

DEFINITION 35. We refer to the parameter λ as the *turning parameter* of the geodesic $\gamma_{0,\lambda}$, and to the point $T(\lambda) = (X(\lambda), Y(\lambda))$ as the *turning point* on the geodesic $\gamma_{0,\lambda}$.

SUMMARY 36. We summarize the turning behaviour of the geodesic $\gamma_{0,\lambda}$ as the turning parameter λ decreases from ∞ to 0:

(1) When $\lambda = \infty$ the geodesic $\gamma_{0,\infty}$ is horizontal,
(2) As λ decreases from ∞ to $f(\infty)$, the geodesics $\gamma_{0,\lambda}$ are asymptotically lines whose slopes increase to infinity,
(3) At $\lambda = f(\infty)$ the geodesic $\gamma_{0,f(\infty)}$ has slope that increases to infinity as x increases,
(4) As λ decreases from $f(\infty)$ to 0, the geodesics $\gamma_{0,\lambda}$ are turn back at $X(\lambda) = f^{-1}(\lambda)$, and return to the y-axis in a path symmetric about the line $y = Y(\lambda)$.

Solving for $\frac{dy}{dx}$ we obtain the equation

$$\frac{dy}{dx} = \frac{\pm f^2(x)}{\sqrt{\lambda^2 - f(x)^2}}.$$

Thus the geodesic $\gamma_{0,\lambda}$ that starts from the origin going in the vertical direction for $x > 0$, and with turning parameter λ, is given by

$$y = \int_0^x \frac{f(u)^2}{\sqrt{\lambda^2 - f(u)^2}} du, \qquad x > 0.$$

Since the metric is invariant under vertical translations, we see that the geodesic $\gamma_{\eta,\lambda}(t)$ whose lower point of intersection with the y-axis has coordinates $(0, \eta)$, and whose positive turning parameter is λ, is given by the equation

$$y = \eta + \int_0^x \frac{f(u)^2}{\sqrt{\lambda^2 - f(u)^2}} du, \qquad x > 0.$$

Thus the entire family of A-geodesics in the right half plane is $\{\gamma_{\eta,\lambda}\}$ parameterized by $(\eta, \lambda) \in (-\infty, \infty) \times (0, \infty]$, where when $\lambda = \infty$, the geodesic $\gamma_{\eta,\infty}(t)$ is the horizontal line through the point $(0, \eta)$.

2. Calculation of A-arc length

Let dt denote A-arc length along the geodesic $\gamma_{0,\lambda}$ and let ds denote Euclidean arc length along $\gamma_{0,\lambda}$.

LEMMA 37. *For $0 < x < X(\lambda)$ and (x, y) on the lower half of the geodesic $\gamma_{0,\lambda}$ we have*

$$\begin{aligned} \frac{dy}{dx} &= \frac{f(x)^2}{\sqrt{\lambda^2 - f(x)^2}}, \\ \frac{dt}{dx}(x,y) &= \frac{\lambda}{\sqrt{\lambda^2 - f(x)^2}}, \\ \frac{dt}{ds}(x,y) &= \frac{\lambda}{\sqrt{\lambda^2 - f(x)^2\left[1 - f(x)^2\right]}}. \end{aligned}$$

PROOF. First we note that $y = \int_0^x \frac{f(u)^2}{\sqrt{\lambda^2 - f^2(u)}} du$ implies $\frac{dy}{dx} = \frac{f(x)^2}{\sqrt{\lambda^2 - f(x)^2}}$. Thus from $dt^2 = dx^2 + \frac{1}{f(x)^2} dy^2$ we have

$$\begin{aligned} \left(\frac{dt}{dx}\right)^2 &= 1 + \frac{1}{f(x)^2}\left(\frac{dy}{dx}\right)^2 = 1 + \frac{1}{f(x)^2}\left(\frac{dy}{dx}\right)^2 \\ &= 1 + \frac{1}{f(x)^2}\frac{f(x)^4}{\lambda^2 - f(x)^2} = \frac{\lambda^2}{\lambda^2 - f(x)^2}. \end{aligned}$$

Then the density of t with respect to s at the point (x, y) on the lower half of the geodesic $\gamma_{0,\lambda}$ is given by

$$\begin{aligned} \frac{dt}{ds} &= \frac{\frac{dt}{dx}}{\frac{ds}{dx}} = \frac{\frac{\lambda}{\sqrt{\lambda^2 - f(x)^2}}}{\sqrt{1 + \left(\frac{dy}{dx}\right)^2}} = \frac{\frac{\lambda}{\sqrt{\lambda^2 - f(x)^2}}}{\sqrt{1 + \frac{f(x)^4}{\lambda^2 - f(x)^2}}} \\ &= \frac{\lambda}{\sqrt{\left(\lambda^2 - f(x)^2\right)\left(1 + \frac{f(x)^4}{\lambda^2 - f(x)^2}\right)}} = \frac{\lambda}{\sqrt{\lambda^2 - f(x)^2 + f(x)^4}} \\ &= \frac{\lambda}{\sqrt{\lambda^2 - f(x)^2\left[1 - f(x)^2\right]}}. \end{aligned}$$

□

Thus at the y-axis when $x = 0$, we have $\frac{dt}{ds} = 1$, and at the turning point $T(\lambda) = (X(\lambda), Y(\lambda))$ of the geodesic, when $\lambda^2 = f(x)^2$, we have $\frac{dt}{ds} = \frac{1}{\lambda} = \frac{1}{f(x)}$. This reflects the fact that near the y axis, the geodesic is nearly horizontal and so the metric arc length is close to Euclidean arc length; while at the turning point for λ small, the density of metric arc length is large compared to Euclidean arc length since movement in the vertical direction meets with much resistance when x is small.

In order to make precise estimates of arc length, we will need to assume some additional properties on the function $f(x)$ when $|x|$ is small.

Assumptions: Fix $R>0$ and let $F(x)=-\ln f(x)$ for $0<x<R$, so that

$$f(x)=e^{-F(|x|)}, \qquad 0<|x|<R.$$

We assume the following for some constants $C\geq 1$ and $0<\varepsilon<1$:

(1) $\lim_{x\to 0^+} F(x)=+\infty$;
(2) $F'(x)<0$ and $F''(x)>0$ for all $x\in(0,R)$;
(3) $\frac{1}{C}|F'(r)|\leq |F'(x)|\leq C|F'(r)|, \qquad \frac{1}{2}r<x<2r<R$;
(4) $\frac{1}{-xF'(x)}$ is increasing in the interval $(0,R)$ and satisfies $\frac{1}{-xF'(x)}\leq\frac{1}{\varepsilon}$ for $x\in(0,R)$;
(5) $\frac{F''(x)}{-F'(x)}\approx\frac{1}{x}$ for $x\in(0,R)$.

These assumptions have the following consequences.

LEMMA 38. *Suppose that R, f and F are as above.*

(1) *If $0<x_1<x_2<R$, then we have*

$$F(x_1)>F(x_2)+\varepsilon\ln\frac{x_2}{x_1}, \text{ equivalently } f(x_1)<\left(\frac{x_1}{x_2}\right)^{\varepsilon} f(x_2).$$

(2) *If $x_1,x_2\in(0,R)$ and* $\max\left\{\varepsilon x_1, x_1-\frac{1}{|F'(x_1)|}\right\}\leq x_2\leq x_1+\frac{1}{|F'(x_1)|}$, *then we have*

$$\begin{aligned}|F'(x_1)| &\approx |F'(x_2)|,\\ f(x_1) &\approx f(x_2).\end{aligned}$$

(3) *If $x\in(0,R)$, then we have*

$$\frac{F''(x)}{|F'(x)|^2}\approx\frac{1}{-xF'(x)}\lesssim 1.$$

PROOF. Assumptions (2) and (4) give $|F'(x_1)|>\frac{\varepsilon}{x}$, and so we have

$$F(x_1)-F(x_2)>\int_{x_1}^{x_2}\frac{\varepsilon}{x}dx=\varepsilon\ln\frac{x_2}{x_1},$$

which proves Part (1) of the lemma. Without loss of generality, assume now that $x_1\leq x_2\leq x_1+\frac{1}{|F'(x_1)|}$. Then by Assumption (4) we also have $x_1\leq x_2\leq\left(1+\frac{1}{\varepsilon}\right)x_1$, and then by Assumption (3), the first assertion in Part (2) of the lemma holds, and with the bound,

$$\begin{aligned}F(x_1)-F(x_2) &= \int_{x_1}^{x_2}-F'(x)\, dx\\ &\approx |F'(x_1)|(x_2-x_1)\leq 1.\end{aligned}$$

From this we get

$$1\leq\frac{f(x_2)}{f(x_1)}=e^{F(x_1)-F(x_2)}\lesssim 1,$$

which proves the second assertion in Part (2) of the lemma. Finally, Assumptions (4) and (5) give

$$\frac{F''(x)}{|F'(x)|^2} = \frac{F''(x)}{-F'(x)} \frac{1}{-F'(x)} \approx \frac{1}{x} \frac{1}{-F'(x)} \lesssim 1,$$

which proves Part (3) of the lemma. □

LEMMA 39. *Suppose $\lambda > 0$, $0 < x < X(\lambda)$ and*

$$y = \int_0^x \frac{f(u)^2}{\sqrt{\lambda^2 - f^2(u)}} du.$$

Then (x, y) lies on the lower half of the geodesic $\gamma_{0,\lambda}$ and

$$y \approx \frac{f(x)^2}{\lambda |F'(x)|}.$$

PROOF. Using first that $\frac{f(u)^2}{\sqrt{\lambda^2 - f^2(u)}}$ is increasing in u, and then that $F(u) = -\ln f(u)$, we have

$$y = \int_0^x \frac{f(u)^2}{\sqrt{\lambda^2 - f(u)^2}} du \approx \int_{\frac{x}{2}}^x \frac{f(u)^2}{\sqrt{\lambda^2 - f(u)^2}} du = \int_{\frac{x}{2}}^x \frac{1}{-2F'(u)} \frac{\left[f(u)^2\right]'}{\sqrt{\lambda^2 - f(u)^2}} du,$$

and then using Assumption (3) we get

$$y \approx \frac{1}{-F'(x)} \int_{\frac{x}{2}}^x \frac{\left[f(u)^2\right]' du}{2\sqrt{\lambda^2 - f(u)^2}} = \frac{1}{-F'(x)} \int_{f(\frac{x}{2})^2}^{f(x)^2} \frac{dv}{2\sqrt{\lambda^2 - v}}.$$

Now from Part (1) of Lemma 38 we obtain $f\left(\frac{x}{2}\right)^2 < \left(\frac{1}{2}\right)^{2\varepsilon} f(x)^2$ and so

$$y \approx \frac{1}{-F'(x)} \int_0^{f(x)^2} \frac{dv}{2\sqrt{\lambda^2 - v}} = \frac{\lambda - \sqrt{\lambda^2 - f(x)^2}}{-F'(x)} \approx \frac{f(x)^2}{\lambda |F'(x)|},$$

where the final estimate follows from $1 - \sqrt{1-t} = \frac{t}{1+\sqrt{1-t}} \approx t$, $0 < t < 1$, with $t = \frac{f(x)^2}{\lambda^2}$. □

REMARK 40. We actually have the upper bound $y \leq \frac{f(x)^2}{\lambda |F'(x)|}$ since $F''(x) > 0$. Indeed, then $\frac{1}{-F'(x)}$ is increasing and for $f(x) < \lambda$ we have

$$\begin{aligned} y &= \int_0^x \frac{f(u)^2}{\sqrt{\lambda^2 - f(u)^2}} du = \int_0^x \frac{1}{-2F'(u)} \frac{\left[f(u)^2\right]'}{\sqrt{\lambda^2 - f(u)^2}} du \\ &\leq \frac{1}{-F'(x)} \int_0^x \frac{\left[f(u)^2\right]'}{2\sqrt{\lambda^2 - f(u)^2}} du = \frac{f(x)^2}{-\lambda F'(x)}. \end{aligned}$$

Now we can estimate the A-arc length of the geodesic $\gamma_{0,\lambda}$ between the two points $P_0 = (0,0)$ and $P_1 = (x_1, y_1)$ where $0 < x_1 < X(\lambda)$ and

$$y_1 = \int_0^{x_1} \frac{f(u)^2}{\sqrt{\lambda^2 - f^2(u)}} du.$$

We have the formula

$$d(P_0, P_1) = \int_{P_0}^{P_1} dt = \int_{P_0}^{P_1} \frac{dt}{dx} dx = \int_0^{x_1} \frac{\lambda}{\sqrt{\lambda^2 - f(x)^2}} dx,$$

from which we obtain $x_1 < d(P_0, P_1)$.

LEMMA 41. *With notation as above we have*

$$\begin{aligned} & x_1 < d(P_0, P_1) \leq d((0,0),(x_1,0)) + d((x_1,0),(x_1,y_1)) \ ; \\ & d((0,0),(x_1,0)) = x_1 \ , \\ & d((x_1,0),(x_1,y_1)) \leq \frac{f(x_1)}{-\lambda F'(x_1)} \leq \frac{1}{-F'(x_1)} < \frac{1}{\varepsilon} x_1 \ . \end{aligned}$$

In particular we have $d(P_0, P_1) \approx x_1$.

PROOF. From Remark 40 we have

$$d((x_1,0),(x_1,y_1)) \leq \frac{y_1}{f(x_1)} \leq \frac{f(x_1)}{-\lambda F'(x_1)},$$

and then we use $f(x_1) \leq \lambda$ and Assumption (4). □

COROLLARY 42. $|F'(d(P_0,P_1))| \approx |F'(x_1)|$ *and* $f(d(P_0,P_1)) \approx f(x_1)$.

PROOF. Combine Part (2) of Lemma 38 with Lemma 41. □

3. Integration over A-balls and Area

Here we investigate properties of the A-ball $B(0, r_0)$ centered at the origin 0 with radius $r_0 > 0$:

$$B(0, r_0) \equiv \left\{ x \in \mathbb{R}^2 : d(0, x) < r_0 \right\}, \qquad r_0 > 0.$$

For this we will use 'A-polar coordinates' where $d(0,(x,y))$ plays the role of the radial variable, and the turning parameter λ plays the role of the angular coordinate. More precisely, given Cartesian coordinates (x,y), the A-polar coordinates (r,λ) are given implicitly by the pair of equations

$$\begin{aligned} r &= \int_0^x \frac{\lambda}{\sqrt{\lambda^2 - f(u)^2}} du \ , \\ y &= \int_0^x \frac{f(u)^2}{\sqrt{\lambda^2 - f(u)^2}} du \ . \end{aligned} \tag{7.1}$$

In this section we will work out the change of variable formula for the quarter A-ball $QB(0, r_0)$ lying in the first quadrant. See Figure 1.

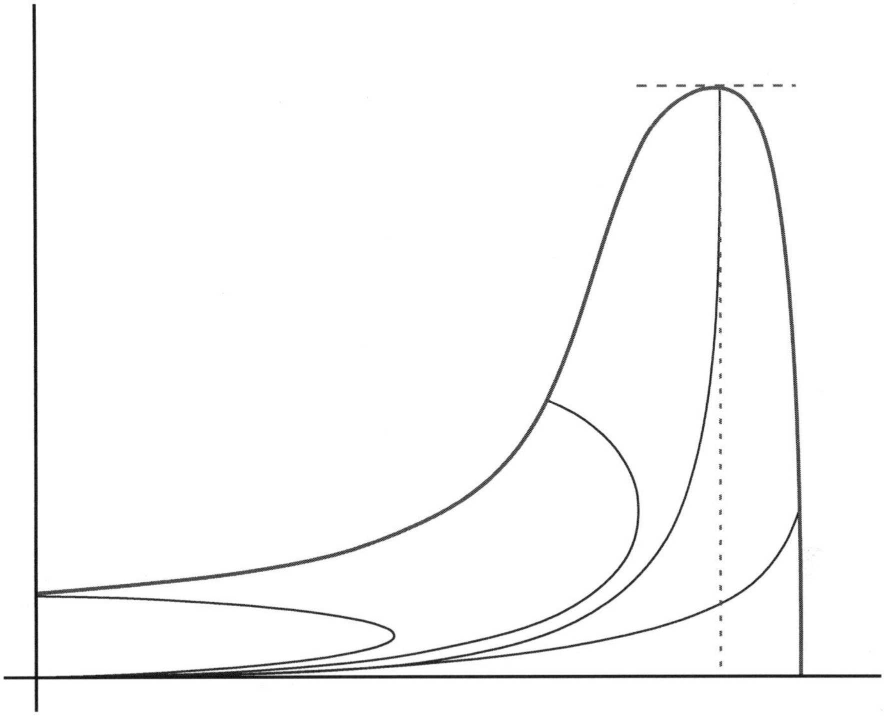

FIGURE 1. A first quadrant view of a control ball centered at the origin.

DEFINITION 43. Let $\lambda \in (0,\infty)$. The geodesic with turning parameter λ first moves to the right and then curls back at the turning point $T(\lambda) = (X(\lambda), Y(\lambda))$ when $x = X(\lambda) \equiv f^{-1}(\lambda)$. If $R(\lambda)$ denotes the A-arc length from the origin to the turning point $T(\lambda)$, we have

$$\begin{aligned} R(\lambda) &\equiv d(0,T(\lambda)) = \int_0^{X(\lambda)} \frac{\lambda}{\sqrt{\lambda^2 - f(u)^2}} du, \\ Y(\lambda) &= \int_0^{X(\lambda)} \frac{f(u)^2}{\sqrt{\lambda^2 - f(u)^2}} du. \end{aligned}$$

The two parts of the geodesic $\gamma_{0,\lambda}$,cut at the point $T(\lambda)$, have different equations:

$$y = \begin{cases} \int_0^x \frac{f(u)^2}{\sqrt{\lambda^2 - f(u)^2}} du & \text{when } y \in [0, Y(\lambda)] \\ 2Y(\lambda) - \int_0^x \frac{f(u)^2}{\sqrt{\lambda^2 - f(u)^2}} du & \text{when } y \in [Y(\lambda), 2Y(\lambda)] \end{cases}. \tag{7.2}$$

We define the region covered by the first equation for the geodesics to be Region 1, and the region covered by the second equation for the geodesics to be Region 2. They are separated by the curve $y = Y(f(x))$. We now calculate the first derivative matrix $\begin{bmatrix} \frac{\partial x}{\partial r} & \frac{\partial x}{\partial \lambda} \\ \frac{\partial y}{\partial r} & \frac{\partial y}{\partial \lambda} \end{bmatrix}$ and the Jacobian $\frac{\partial(x,y)}{\partial(r,\lambda)}$ in Regions 1 and 2 separately.

3.1. Region 1. Applying implicit differentiation to the first equation in (7.1), we have

$$1 = \frac{\partial r}{\partial r} = \frac{\partial x}{\partial r} \cdot \frac{\lambda}{\sqrt{\lambda^2 - f(x)^2}},$$

$$0 = \frac{\partial r}{\partial \lambda} = \frac{\partial x}{\partial \lambda} \cdot \frac{\lambda}{\sqrt{\lambda^2 - f(x)^2}} + \int_0^x \frac{\partial}{\partial \lambda}\left[\frac{\lambda}{\sqrt{\lambda^2 - f(u)^2}}\right] du,$$

where

$$\frac{\partial}{\partial \lambda}\left[\frac{\lambda}{\sqrt{\lambda^2 - f(u)^2}}\right] = \frac{1 \cdot \sqrt{\lambda^2 - f(u)^2} - \lambda \cdot \frac{2\lambda}{2\sqrt{\lambda^2 - f(u)^2}}}{\lambda^2 - f(u)^2} = \frac{-f(u)^2}{\left(\lambda^2 - f(u)^2\right)^{\frac{3}{2}}}.$$

Thus we have

$$\begin{aligned} \frac{\partial x}{\partial r} &= \frac{\sqrt{\lambda^2 - f(x)^2}}{\lambda}, \\ \frac{\partial x}{\partial \lambda} &= \frac{\sqrt{\lambda^2 - f(x)^2}}{\lambda} \cdot \int_0^x \frac{f(u)^2}{\left(\lambda^2 - f(u)^2\right)^{\frac{3}{2}}} du. \end{aligned}$$

Applying implicit differentiation to the second equation in (7.1), we have

$$\frac{\partial y}{\partial r} = \frac{\partial x}{\partial r} \cdot \frac{f(x)^2}{\sqrt{\lambda^2 - f(x)^2}};$$

$$\begin{aligned} \frac{\partial y}{\partial \lambda} &= \frac{\partial x}{\partial \lambda} \cdot \frac{f(x)^2}{\sqrt{\lambda^2 - f(x)^2}} + \int_0^x \frac{\partial}{\partial \lambda}\left[\frac{f(u)^2}{\sqrt{\lambda^2 - f(u)^2}}\right] du \\ &= \frac{\partial x}{\partial \lambda} \cdot \frac{f(x)^2}{\sqrt{\lambda^2 - f(x)^2}} + \int_0^x \frac{-\lambda f(u)^2}{\left(\lambda^2 - f(u)^2\right)^{\frac{3}{2}}} du \end{aligned}$$

Plugging the equation for $\frac{\partial x}{\partial \lambda}$ into these equations, we obtain

$$\begin{aligned} \frac{\partial y}{\partial r} &= \frac{f(x)^2}{\lambda}, \\ \frac{\partial y}{\partial \lambda} &= \frac{f(x)^2 - \lambda^2}{\lambda} \cdot \int_0^x \frac{f(u)^2}{\left(\lambda^2 - f(u)^2\right)^{\frac{3}{2}}} du, \end{aligned}$$

and this completes the calculation of the first derivative matrix $\begin{bmatrix} \frac{\partial x}{\partial r} & \frac{\partial x}{\partial \lambda} \\ \frac{\partial y}{\partial r} & \frac{\partial y}{\partial \lambda} \end{bmatrix}$.

Now we can calculate the Jacobian

$$\begin{aligned}\frac{\partial(x,y)}{\partial(r,\lambda)} &= \det\begin{bmatrix}\frac{\sqrt{\lambda^2-f(x)^2}}{\lambda} & \frac{\sqrt{\lambda^2-f(x)^2}}{\lambda}\cdot\int_0^x \frac{f(u)^2}{\left(\lambda^2-f(u)^2\right)^{\frac{3}{2}}}du \\ \frac{f(x)^2}{\lambda} & \frac{f(x)^2-\lambda^2}{\lambda}\cdot\int_0^x \frac{f(u)^2}{\left(\lambda^2-f(u)^2\right)^{\frac{3}{2}}}du\end{bmatrix} \\ &= -\sqrt{\lambda^2-f(x)^2}\int_0^x \frac{f(u)^2}{\left(\lambda^2-f(u)^2\right)^{\frac{3}{2}}}du.\end{aligned}$$

In addition we have

$$\int_0^x \frac{f(u)^2}{\left(\lambda^2-f(u)^2\right)^{\frac{3}{2}}}du \approx \int_{x/2}^x \frac{f(u)^2}{\left(\lambda^2-f(u)^2\right)^{\frac{3}{2}}}du = \int_{x/2}^x \frac{\frac{d}{du}\left[f(u)^2\right]\cdot\frac{f(u)}{2f'(u)}}{\left(\lambda^2-f(u)^2\right)^{\frac{3}{2}}}du,$$

where

$$\frac{f(u)}{2f'(u)} = \frac{1}{-2F'(u)} \approx \frac{1}{-F'(x)},$$

and so we have

$$\int_0^x \frac{f(u)^2}{\left(\lambda^2-f(u)^2\right)^{\frac{3}{2}}}du \approx \frac{1}{-F'(x)}\int_{f\left(\frac{x}{2}\right)^2}^{f(x)^2} \frac{1}{\left(\lambda^2-v\right)^{\frac{3}{2}}}dv.$$

By Part (1) of Lemma 38, we have $f\left(\frac{x}{2}\right) < \left(\frac{1}{2}\right)^{\varepsilon} f(x)$, and as a result, we obtain

$$\begin{aligned}\int_0^x \frac{f(u)^2}{\left(\lambda^2-f(u)^2\right)^{\frac{3}{2}}}du &\approx \frac{1}{-F'(x)}\int_0^{f(x)^2} \frac{1}{\left(\lambda^2-v\right)^{\frac{3}{2}}}dv \\ &\approx \frac{1}{-F'(x)}\left(\frac{1}{\sqrt{\lambda^2-f(x)^2}}-\frac{1}{\lambda}\right).\end{aligned}$$

Altogether we have the estimate

$$\left|\frac{\partial(x,y)}{\partial(r,\lambda)}\right| \approx \frac{1}{-F'(x)}\cdot\frac{\lambda-\sqrt{\lambda^2-f(x)^2}}{\lambda} \approx \frac{f(x)^2}{\lambda^2\left|F'(x)\right|}. \tag{7.3}$$

From Corollary 42, we also have

$$\left|\frac{\partial(x,y)}{\partial(r,\lambda)}\right| \approx \frac{f(r)^2}{\lambda^2\left|F'(r)\right|}. \tag{7.4}$$

3.2. Region 2. In Region 2 we have the following pair of formulas:

$$\begin{aligned} r &= 2R(\lambda) - \int_0^x \frac{\lambda}{\sqrt{\lambda^2-f(u)^2}}du\ , \\ y &= 2Y(\lambda) - \int_0^x \frac{f(u)^2}{\sqrt{\lambda^2-f(u)^2}}du\ . \end{aligned} \tag{7.5}$$

where we recall that $R(\lambda) = \int_0^{X(\lambda)} \frac{\lambda}{\sqrt{\lambda^2-f(u)^2}}du$ is the arc length of the geodesic $\gamma_{0,\lambda}$ from the origin 0 to the turning point $T(\lambda)$. Before proceeding, we calculate the derivative of $Y(\lambda)$. We note that due to cancellation, the derivative $R'(\lambda)$ does not explicitly enter into the formula for the Jacobian $\frac{\partial(x,y)}{\partial(r,\lambda)}$ below, so we defer its calculation for now.

LEMMA 44. *The derivative of* $Y(\lambda)$ *is given by*

$$Y'(\lambda) = \int_0^{f^{-1}(\lambda)} \frac{F''(u)}{|F'(u)|^2} \frac{\lambda}{\sqrt{\lambda^2 - f(u)^2}} du.$$

PROOF. Integrating by parts we obtain

$$\begin{aligned} Y(\lambda) &= \int_0^{f^{-1}(\lambda)} \frac{-f(u)}{f'(u)} \cdot \frac{d}{du} \sqrt{\lambda^2 - f(u)^2} du \\ &= \int_0^{f^{-1}(\lambda)} \frac{1}{F'(u)} \cdot \frac{d}{du} \sqrt{\lambda^2 - f(u)^2} du \\ &= -\int_0^{f^{-1}(\lambda)} \sqrt{\lambda^2 - f(u)^2} \cdot \frac{d}{du} \frac{1}{F'(u)} du \\ &= \int_0^{f^{-1}(\lambda)} \frac{F''(u)}{|F'(u)|^2} \sqrt{\lambda^2 - f(u)^2} du, \end{aligned}$$

and so from $\lambda^2 - f\left(f^{-1}(\lambda)\right)^2 = 0$, we have

$$Y'(\lambda) = 0 + \int_0^{f^{-1}(\lambda)} \frac{F''(u)}{|F'(u)|^2} \frac{\lambda}{\sqrt{\lambda^2 - f(u)^2}} du.$$

□

Applying implicit differentiation to the first equation in (7.5), we have

$$1 = \frac{\partial r}{\partial r} = -\frac{\partial x}{\partial r} \cdot \frac{\lambda}{\sqrt{\lambda^2 - f(x)^2}},$$

$$0 = \frac{\partial r}{\partial \lambda} = 2R'(\lambda) - \frac{\partial x}{\partial \lambda} \cdot \frac{\lambda}{\sqrt{\lambda^2 - f(x)^2}} - \int_0^x \frac{\partial}{\partial \lambda} \left[\frac{\lambda}{\sqrt{\lambda^2 - f(u)^2}} \right] du,$$

where

$$\frac{\partial}{\partial \lambda} \left[\frac{\lambda}{\sqrt{\lambda^2 - f(u)^2}} \right] = \frac{1 \cdot \sqrt{\lambda^2 - f(u)^2} - \lambda \cdot \frac{2\lambda}{2\sqrt{\lambda^2 - f(u)^2}}}{\lambda^2 - f(u)^2} = \frac{-f(u)^2}{\left(\lambda^2 - f(u)^2\right)^{\frac{3}{2}}}.$$

Thus we have

$$\begin{aligned} \frac{\partial x}{\partial r} &= -\frac{\sqrt{\lambda^2 - f(x)^2}}{\lambda}, \\ \frac{\partial x}{\partial \lambda} &= \frac{2\sqrt{\lambda^2 - f(x)^2}}{\lambda} L'(\lambda) + \frac{\sqrt{\lambda^2 - f(x)^2}}{\lambda} \cdot \int_0^x \frac{f(u)^2}{\left(\lambda^2 - f(u)^2\right)^{\frac{3}{2}}} du. \end{aligned}$$

Applying implicit differentiation to the second equation in (7.5), we have

$$\frac{\partial y}{\partial r} = -\frac{\partial x}{\partial r}\cdot\frac{f(x)^2}{\sqrt{\lambda^2 - f(x)^2}};$$

$$\begin{aligned}\frac{\partial y}{\partial \lambda} &= 2Y'(\lambda) - \frac{\partial x}{\partial \lambda}\cdot\frac{f(x)^2}{\sqrt{\lambda^2 - f(x)^2}} - \int_0^x \frac{\partial}{\partial\lambda}\left[\frac{f(u)^2}{\sqrt{\lambda^2 - f(u)^2}}\right]du \\ &= 2Y'(\lambda) - \frac{\partial x}{\partial \lambda}\cdot\frac{f(x)^2}{\sqrt{\lambda^2 - f(x)^2}} - \int_0^x \frac{-\lambda f(u)^2}{\left(\lambda^2 - f(u)^2\right)^{\frac{3}{2}}}du\end{aligned}$$

Plugging the equation for $\frac{\partial x}{\partial \lambda}$ above into these equations, we have

$$\begin{aligned}\frac{\partial y}{\partial r} &= \frac{f(x)^2}{\lambda}, \\ \frac{\partial y}{\partial \lambda} &= 2Y'(\lambda) - \frac{2f(x)^2}{\lambda}R'(\lambda) + \frac{\lambda^2 - f(x)^2}{\lambda}\cdot\int_0^x \frac{f(u)^2}{\left(\lambda^2 - f(u)^2\right)^{\frac{3}{2}}}du.\end{aligned}$$

Thus the Jacobian is given by

$$\begin{aligned}&\frac{\partial(x,y)}{\partial(r,\lambda)} \\ &= \det\begin{bmatrix} -\frac{\sqrt{\lambda^2-f(x)^2}}{\lambda} & \frac{2\sqrt{\lambda^2-f(x)^2}}{\lambda}R'(\lambda) + \frac{\sqrt{\lambda^2-f(x)^2}}{\lambda}\cdot\int_0^x \frac{f(u)^2}{(\lambda^2-f(u)^2)^{\frac{3}{2}}}du \\ \frac{f(x)^2}{\lambda} & 2Y'(\lambda) - \frac{2f(x)^2}{\lambda}R'(\lambda) + \frac{\lambda^2-f(x)^2}{\lambda}\cdot\int_0^x \frac{f(u)^2}{(\lambda^2-f(u)^2)^{\frac{3}{2}}}du \end{bmatrix} \\ &= \det\begin{bmatrix} -\frac{\sqrt{\lambda^2-f(x)^2}}{\lambda} & \frac{\sqrt{\lambda^2-f(x)^2}}{\lambda}\cdot\int_0^x \frac{f(u)^2}{(\lambda^2-f(u)^2)^{\frac{3}{2}}}du \\ \frac{f(x)^2}{\lambda} & 2Y'(\lambda) + \frac{\lambda^2-f(x)^2}{\lambda}\cdot\int_0^x \frac{f(u)^2}{(\lambda^2-f(u)^2)^{\frac{3}{2}}}du \end{bmatrix} \\ &= -\sqrt{\lambda^2 - f(x)^2}\left\{\int_0^x \frac{f(u)^2}{\left(\lambda^2 - f(u)^2\right)^{\frac{3}{2}}}du + \frac{2}{\lambda}Y'(\lambda)\right\} \\ &= -\sqrt{\lambda^2 - f(x)^2}\left\{\int_0^x \frac{f(u)^2}{\left(\lambda^2 - f(u)^2\right)^{\frac{3}{2}}}du + \frac{2}{\lambda}\int_0^{f^{-1}(\lambda)} \frac{F''(u)}{|F'(u)|^2}\frac{\lambda}{\sqrt{\lambda^2 - f(u)^2}}du\right\} \\ &= -\sqrt{\lambda^2 - f(x)^2}\left\{\int_0^x \frac{f(u)^2}{\left(\lambda^2 - f(u)^2\right)^{\frac{3}{2}}}du + \int_0^{f^{-1}(\lambda)} \frac{F''(u)}{|F'(u)|^2}\cdot\frac{2}{\sqrt{\lambda^2 - f(u)^2}}du\right\}.\end{aligned}$$

In fact, we have

$$\begin{aligned}\int_0^x \frac{f(u)^2}{\left(\lambda^2 - f(u)^2\right)^{\frac{3}{2}}} du &= \int_0^x \frac{f(u)}{f'(u)} \cdot \frac{d}{du}\left[\frac{1}{\sqrt{\lambda^2 - f(u)^2}}\right] du \\ &= \int_0^x \frac{1}{-F'(u)} \cdot \frac{d}{du}\left[\frac{1}{\sqrt{\lambda^2 - f(u)^2}}\right] du \\ &= \frac{1}{-F'(x)} \cdot \frac{1}{\sqrt{\lambda^2 - f(x)^2}} - \int_0^x \frac{1}{\sqrt{\lambda^2 - f(u)^2}} \cdot \frac{d}{du}\left[\frac{1}{-F'(u)}\right] du \\ &= \frac{1}{-F'(x)} \cdot \frac{1}{\sqrt{\lambda^2 - f(x)^2}} - \int_0^x \frac{1}{\sqrt{\lambda^2 - f(u)^2}} \cdot \frac{F''(u)}{|F'(u)|^2} du\end{aligned}$$

As a result, we have within a factor of 2,

$$\begin{aligned}\left|\frac{\partial(x,y)}{\partial(r,\lambda)}\right| &\approx \sqrt{\lambda^2 - f(x)^2}\left\{\frac{1}{-F'(x)} \cdot \frac{1}{\sqrt{\lambda^2 - f(x)^2}}\right. \\ &\qquad \left. + \int_0^{f^{-1}(\lambda)} \frac{F''(u)}{|F'(u)|^2} \cdot \frac{1}{\sqrt{\lambda^2 - f(u)^2}} du\right\} \\ &= \frac{1}{-F'(x)} + \sqrt{\lambda^2 - f(x)^2} \int_0^{f^{-1}(\lambda)} \frac{F''(u)}{|F'(u)|^2} \cdot \frac{1}{\sqrt{\lambda^2 - f(u)^2}} du.\end{aligned} \tag{7.6}$$

By Assumption (5), we have

$$\int_0^{f^{-1}(\lambda)} \frac{F''(u)}{|F'(u)|^2} \cdot \frac{1}{\sqrt{\lambda^2 - f(u)^2}} du \approx \int_0^{f^{-1}(\lambda)} \frac{1}{-uF'(u)} \cdot \frac{1}{\sqrt{\lambda^2 - f(u)^2}} du.$$

By Assumptions (3) and (4), the function $\frac{1}{-uF'(u)}$ increases and satisfies the doubling property, and so

$$\begin{aligned}&\int_0^{f^{-1}(\lambda)} \frac{F''(u)}{|F'(u)|^2} \cdot \frac{1}{\sqrt{\lambda^2 - f(u)^2}} du \\ &\qquad \approx \frac{1}{-f^{-1}(\lambda) F'(f^{-1}(\lambda))} \int_0^{f^{-1}(\lambda)} \frac{1}{\sqrt{\lambda^2 - f(u)^2}} du \\ &\qquad = \frac{1}{-f^{-1}(\lambda) F'(f^{-1}(\lambda))} \frac{R(\lambda)}{\lambda} \\ &\qquad \simeq \frac{1}{-\lambda F'(f^{-1}(\lambda))}\end{aligned} \tag{7.7}$$

since $R(\lambda) \approx f^{-1}(\lambda)$ by Lemma 41. Finally we can combine (7.6) and (7.7) to obtain

$$\left|\frac{\partial(x,y)}{\partial(r,\lambda)}\right| \approx \frac{1}{-F'(x)} + \frac{\sqrt{\lambda^2 - f(x)^2}}{\lambda} \cdot \frac{1}{-F'(f^{-1}(\lambda))} \approx \frac{1}{-F'(f^{-1}(\lambda))}.$$

According to Corollary 42, we also have

$$\left|\frac{\partial(x,y)}{\partial(r,\lambda)}\right| \approx \frac{1}{-F'(R(\lambda))}.$$

3.3. Integral of Radial Functions. Summarizing our estimates on the Jacobian we have

$$\left|\frac{\partial(x,y)}{\partial(r,\lambda)}\right| \approx \begin{cases} \frac{f(r)^2}{\lambda^2|F'(r)|} \simeq \frac{f(x)^2}{\lambda^2|F'(x)|} & \text{when } r < R(\lambda) \\ \frac{1}{|F'(f^{-1}(\lambda))|} \simeq \frac{1}{|F'(R(\lambda))|} & \text{when } R(\lambda) < r < 2R(\lambda) \end{cases}.$$

Therefore we have the following change of variable formula for nonnegative functions w:

$$\iint_{QB(0,r_0)} w\,dxdy = \int_0^{r_0} \left[\int_{R^{-1}\left(\frac{r}{2}\right)}^{\infty} w \left|\frac{\partial(x,y)}{\partial(r,\lambda)}\right| d\lambda\right] dr$$

$$\approx \int_0^{r_0} \left[\int_{R^{-1}\left(\frac{r}{2}\right)}^{R^{-1}(r)} w(r,\lambda) \frac{1}{|F'(R(\lambda))|} d\lambda + \int_{R^{-1}(r)}^{\infty} w(r,\lambda) \frac{f(r)^2}{\lambda^2 |F'(r)|} d\lambda\right] dr$$

$$\approx \int_0^{r_0} \left[\int_{R^{-1}\left(\frac{r}{2}\right)}^{R^{-1}(r)} w(r,\lambda) \frac{1}{|F'(r)|} d\lambda + \int_{R^{-1}(r)}^{\infty} w(r,\lambda) \frac{f^2(r)}{\lambda^2|F'(r)|} d\lambda\right] dr$$

If w is a radial function, then we have

$$\iint_{QB(0,r_0)} w\,dxdy \approx \int_0^{r_0} w(r) \left[\int_{R^{-1}\left(\frac{r}{2}\right)}^{R^{-1}(r)} \frac{1}{|F'(r)|} d\lambda + \int_{R^{-1}(r)}^{\infty} \frac{f(r)^2}{\lambda^2 |F'(r)|} d\lambda\right] dr$$

$$\approx \int_0^{r_0} w(r) \left[\frac{R^{-1}(r) - R^{-1}\left(\frac{r}{2}\right)}{|F'(r)|} + \frac{f(r)^2}{R^{-1}(r)\,|F'(r)|}\right] dr.$$

From Corollary 42, we have $R^{-1}(r) \simeq f(r)$, and so we have

$$\iint_{B(0,r_0)} w(r)\,dxdy \approx \int_0^{r_0} w(r) \frac{f(r)}{|F'(r)|} dr. \tag{7.8}$$

CONCLUSION 45. The area of the A-ball $B(0,r_0)$ satisfies

$$|B(0,r_0)| = \iint_{B(0,r_0)} dxdy \approx \int_0^{r_0} \frac{f(r)}{|F'(r)|} dr \approx \frac{f(r_0)}{|F'(r_0)|^2}. \tag{7.9}$$

PROOF. Since $F(r) = -\ln f(r)$, we have $F'(r) = -\frac{f'(r)}{f(r)}$ and $\frac{f(r)}{-F'(r)} = \frac{f(r)^2}{f'(r)} = \frac{f(r)^2}{f'(r)^2} f'(r) = \frac{f'(r)}{|F'(r)|^2}$, and so

$$\iint_{B(0,r_0)} dxdy \approx \int_0^{r_0} \frac{f(r)}{|F'(r)|} dr \approx \int_{\frac{r_0}{2}}^{r_0} \frac{f(r)}{|F'(r)|} dr = \int_{\frac{r_0}{2}}^{r_0} \frac{f'(r)}{|F'(r)|^2} dr$$

$$\approx \frac{1}{|F'(r_0)|^2} \int_{\frac{r_0}{2}}^{r_0} f'(r)\,dr = \frac{f(r_0) - f\left(\frac{r_0}{2}\right)}{|F'(r_0)|^2} \approx \frac{f(r_0)}{|F'(r_0)|^2}.$$

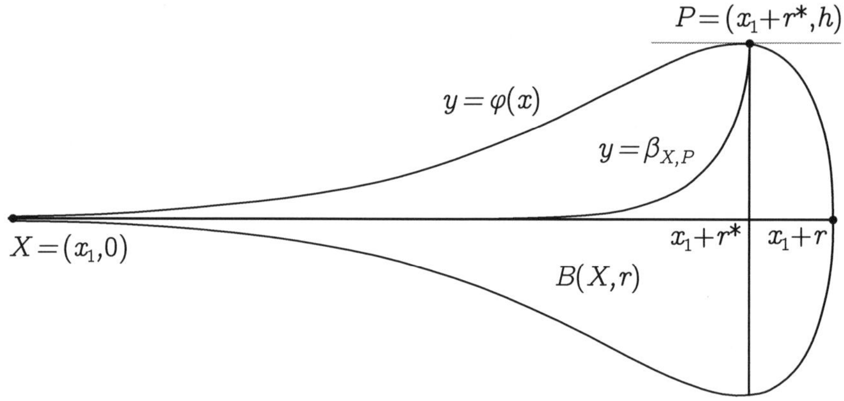

FIGURE 2. The right 'half' of a control ball centered at $(x_1, 0)$.

□

3.4. Balls centered at an arbitrary point. In this section we consider the "height" of an arbitrary A-ball and the relative position at which it is achieved in the ball.

DEFINITION 46. Let $X = (x_1, 0)$ be a point on the positive x-axis and let r be a positive real number. Let the upper half of the boundary of the ball $B(X, r)$ be given as a graph of the function $\varphi(x)$, $x_1 - r < x < x_1 + r$. Denote by $\beta_{X,Q}$ the geodesic connecting the center X of the ball $B(X, r)$ with a point Q on the boundary $\partial B(X, r)$ of the ball $B(X, r)$.

Denote by $P = P_{x_1,r} = (x_1 + r^*, h)$ the unique point on the boundary $\partial B(X, r)$ of the ball $B(X, r)$ with $r^* > 0$ and $h > 0$ at which the geodesic $\beta_{X,P}$ connecting X and P has a *vertical tangent* at P. This defines

$$r^* = r^*(x_1, r) \text{ and } h = h(x_1, r) = \varphi(x_1 + r^*)$$

implicitly as functions of the two independent variables x_1 and r.

We will often write simply r^* and h in place of $r^*(x_1, r)$ and $h(x_1, r)$ respectively when x_1 and r are understood. See figure 2.

PROPOSITION 47. *Let $\beta_{X,P}$, r^* and h be defined as above. Define $\lambda(x)$ implicitly by*

$$r = \int_{x_1}^{x} \frac{\lambda(x)}{\sqrt{\lambda(x)^2 - f(u)^2}} du.$$

Then

(1) *For $x_1 - r < x < x_1 + r$ we have $\varphi(x) \leq \varphi(x_1 + r^*) = h$.*
(2) *If $r \geq \frac{1}{|F'(x_1)|}$, then*

$$h \approx \frac{f(x_1 + r)}{|F'(x_1 + r)|} \text{ and } r - r^* \approx \frac{1}{|F'(x_1 + r)|}.$$

(3) *If $r \leq \frac{1}{|F'(x_1)|}$, then*

$$h \approx r f(x_1) \text{ and } r - r^* \approx r.$$

We begin by proving part (1) of Proposition 47. First consider the case $x \geq x_1 + r^*$. Then we are in Region 1 and so $\lambda(x) \geq f(x)$ and we have

$$\varphi(x) = \int_{x_1}^{x} \frac{f(u)^2}{\sqrt{\lambda(x)^2 - f(u)^2}} du.$$

Differentiating $\varphi(x)$ we get

$$\varphi'(x) = \frac{f(x)^2}{\sqrt{\lambda(x)^2 - f(x)^2}} - \left(\int_{x_1}^{x} \frac{f(u)^2}{\left(\lambda(x)^2 - f(u)^2\right)^{\frac{3}{2}}} du \right) \lambda(x) \lambda'(x),$$

and differentiating the definition of $\lambda(x)$ implicitly gives

$$0 = \frac{\lambda(x)}{\sqrt{\lambda(x)^2 - f(x)^2}} - \left(\int_{x_1}^{x} \frac{f(u)^2}{\left(\lambda(x)^2 - f(u)^2\right)^{\frac{3}{2}}} du \right) \lambda'(x).$$

Combining equalities yields

$$\varphi'(x) = \frac{f(x)^2}{\sqrt{\lambda(x)^2 - f(x)^2}} - \lambda(x) \frac{\lambda(x)}{\sqrt{\lambda(x)^2 - f(x)^2}} = -\sqrt{\lambda(x)^2 - f(x)^2}.$$

When $x = x_1 + r^*$ we have $\infty = \frac{dy}{dx} = \frac{f(x)^2}{\sqrt{\lambda(x)^2 - f(x)^2}}$, which implies $\lambda(x) = f(x)$, and hence $\varphi'(x_1 + r^*) = 0$. Thus we have $\varphi(x) \leq \varphi(x_1 + r^*) = h$ for $x \geq x_1 + r^*$. Similar arguments show that $\varphi(x) \leq \varphi(x_1 + r^*) = h$ for $x_1 - r \leq x < x_1 + r^*$, and this completes the proof of part (1).

Now we turn to the proofs of parts (2) and (3) of Proposition 47. The locus (x, y) of the geodesic $\beta_{X,P}$ satisfies

$$y = \int_{x_1}^{x} \frac{f(u)^2}{\sqrt{(\lambda^*)^2 - f(u)^2}} du, \tag{7.10}$$

where $\lambda^* = f(x_1 + r^*)$. We will use the following two lemmas in the proofs of parts (2) and (3) of Proposition 47.

LEMMA 48. *The height $h = h(x_1, r)$ and the horizontal displacement $r - r^* = r - r^*(x_1, r)$ satisfy*

$$f(x_1 + r^*) \cdot (r - r^*) \leq h \leq 2f(x_1 + r^*) \cdot (r - r^*).$$

PROOF. The A-arc length r of $\beta_{X,P}$ is given by

$$r = \int_{x_1}^{x_1 + r^*} \frac{\lambda^*}{\sqrt{(\lambda^*)^2 - f(u)^2}} du.$$

Thus

$$\begin{aligned} r-r^* &= \int_{x_1}^{x_1+r^*}\left(\frac{\lambda^*}{\sqrt{(\lambda^*)^2-f(u)^2}}-1\right)du \\ &= \int_{x_1}^{x_1+r^*}\frac{f(u)^2}{\sqrt{(\lambda^*)^2-f(u)^2}}\cdot\frac{1}{\lambda^*+\sqrt{(\lambda^*)^2-f(u)^2}}du \end{aligned}$$

Comparing this with the height $h=\int_{x_1}^{x_1+r^*}\frac{f^2(u)}{\sqrt{(\lambda^*)^2-f(u)^2}}du$, we have

$$\frac{h}{2\lambda^*}\leq r-r^*\leq\frac{h}{\lambda^*}.$$

This completes the proof since $\lambda^*=f(x_1+r^*)$. □

LEMMA 49. *The height h satisfies the estimate*

$$h\approx\frac{1}{|F'(x_1+r^*)|}\sqrt{f(x_1+r^*)^2-f(x_1)^2}.$$

In fact the right hand side is an exact upper bound:

$$h\leq\frac{1}{|F'(x_1+r^*)|}\sqrt{f(x_1+r^*)^2-f(x_1)^2}.$$

PROOF. Using the fact that $\frac{1}{-F'(u)}=\frac{1}{|F'(u)|}$ is increasing, together with the equation (7.10) for the geodesic $\beta_{X,P}$, we have

$$\begin{aligned} h(x_1,r) &= \int_{x_1}^{x_1+r^*(x_1,r)}\frac{f(u)^2}{\sqrt{(\lambda^*)^2-f(u)^2}}du \\ &= \int_{x_1}^{x_1+r^*}\frac{\frac{d}{du}\left[f(u)^2\right]}{\sqrt{(\lambda^*)^2-f(u)^2}}\cdot\frac{1}{-2F'(u)}du \\ &\leq \frac{1}{|F'(x_1+r^*)|}\int_{x_1}^{x_1+r^*}\frac{\frac{d}{du}\left[f(u)^2\right]}{2\sqrt{(\lambda^*)^2-f(u)^2}}du \\ &= \frac{1}{|F'(x_1+r^*)|}\sqrt{f(x_1+r^*)^2-f(x_1)^2}, \end{aligned}$$

where in the last line we used $\lambda^*=f(x_1+r^*)$. To prove the reverse estimate, we consider two cases:

Case 1: If $r^* < x_1$, then we use our assumption that $\frac{1}{-F'(u)} = \frac{1}{|F'(u)|}$ has the doubling property to obtain

$$
\begin{aligned}
h = \int_{x_1}^{x_1+r^*} \frac{f(u)^2}{\sqrt{(\lambda^*)^2 - f(u)^2}} du &= \int_{x_1}^{x_1+r^*} \frac{\frac{d}{du}\left[f(u)^2\right]}{\sqrt{(\lambda^*)^2 - f(u)^2}} \cdot \frac{1}{-2F'(u)} du \\
&\simeq \frac{1}{|F'(x_1+r^*)|} \int_{x_1}^{x_1+r^*} \frac{\frac{d}{du}\left[f(u)^2\right]}{2\sqrt{(\lambda^*)^2 - f(u)^2}} du \\
&= \frac{1}{|F'(x_1+r^*)|} \sqrt{f(x_1+r^*)^2 - f(x_1)^2}.
\end{aligned}
$$

Case 2: If $r^* \geq x_1$, we make a similar estimate by modifying the lower limit of integral, and using the fact that $f(u)$ increases:

$$
\begin{aligned}
h \approx \int_{x_1+\frac{r^*}{2}}^{x_1+r^*} \frac{f(u)^2}{\sqrt{(\lambda^*)^2 - f(u)^2}} du &= \int_{x_1+\frac{r^*}{2}}^{x_1+r^*} \frac{\frac{d}{du}\left[f(u)^2\right]}{\sqrt{(\lambda^*)^2 - f(u)^2}} \cdot \frac{1}{-2F'(u)} du \\
&\approx \frac{1}{|F'(x_1+r^*)|} \int_{x_1+\frac{r^*}{2}}^{x_1+r^*} \frac{\frac{d}{du}\left[f(u)^2\right]}{2\sqrt{(\lambda^*)^2 - f(u)^2}} du \\
&= \frac{1}{|F'(x_1+r^*)|} \sqrt{f(x_1+r^*)^2 - f\left(x_1+\frac{r^*}{2}\right)^2}.
\end{aligned}
$$

Finally we have

$$
\sqrt{f(x_1+r^*)^2 - f\left(x_1+\frac{r^*}{2}\right)^2} \approx f(x_1+r^*) \approx \sqrt{f(x_1+r^*)^2 - f(x_1)^2}
$$

by the assumption $r^* \geq x_1$ together with Part 1 of Lemma 38.

This completes the proof of Lemma 49. □

Corollary 50. *Combining Lemmas* 48 *and* 49, *for* $h = h(x_1, r)$ *and* $r^* = r^*(x_1, r)$, *we have*

$$
f(x_1+r^*) \cdot (r - r^*) \leq h \leq \frac{1}{|F'(x_1+r^*)|} \sqrt{f(x_1+r^*)^2 - f(x_1)^2},
$$

and as a result,

$$
r - r^* \leq \frac{1}{|F'(x_1+r^*)|} \cdot \frac{\sqrt{f(x_1+r^*)^2 - f(x_1)^2}}{f(x_1+r^*)} \leq \frac{1}{|F'(x_1+r^*)|}. \tag{7.11}
$$

From part (2) of Lemma 38 *we obtain*

$$
\begin{aligned}
|F'(x_1+r)| &\approx |F'(x_1+r^*)|, \\
f(x_1+r) &\simeq f(x_1+r^*).
\end{aligned} \tag{7.12}
$$

We now split the proof of Proposition 47 into two cases.

3.4.1. *Proof of part (2) of Proposition* 47 *for* $r \geq \frac{1}{|F'(x_1)|}$. By Lemmas 48 and 49, we have

$$r - r^*(x_1, r) = r - r^* \approx \frac{1}{|F'(x_1 + r^*)|} \cdot \frac{\sqrt{f(x_1 + r^*)^2 - f(x_1)^2}}{f(x_1 + r^*)}. \tag{7.13}$$

We consider two cases.

Case A: If $r^* > r_1 \equiv \frac{1}{2|F'(x_1)|}$, then we have

$$\begin{aligned} F(x_1) - F(x_1 + r^*) &= \int_{x_1}^{x_1 + r^*} |F'(x_1)|\, dx \geq \int_{x_1}^{x_1 + r_1} |F'(x_1)|\, dx \\ &\geq |F'(x_1 + r_1)| \cdot r_1 \gtrsim 1. \end{aligned}$$

Here we used the estimate $|F'(x_1 + r_1)| \approx |F'(x_1)|$ given by Part 2 of Lemma 38. This implies

$$\ln \frac{f(x_1 + r^*)}{f(x_1)} \gtrsim 1,$$

and we have

$$\frac{\sqrt{f(x_1 + r^*)^2 - f(x_1)^2}}{f(x_1 + r^*)} \approx 1.$$

Plugging this into (7.13), we have $r - r^* \approx \frac{1}{|F'(x_1 + r^*)|}$. The proof is completed using (7.12) and Lemma 48.

Case B: If $r^* \leq r_1$, then we have $|F'(x_1 + r^*)| \approx |F'(x_1)|$ and $r - r^* \geq \frac{1}{2|F'(x_1)|}$. Therefore we have

$$r - r^* \gtrsim \frac{1}{|F'(x_1 + r^*)|}.$$

Combining this with (7.11), we obtain $r - r^* \approx \frac{1}{|F'(x_1 + r^*)|}$ again, and the proof is completed as in the first case.

3.4.2. *Proof of part (3) of Proposition* 47 *for* $r \leq \frac{1}{|F'(x_1)|}$. In this case Lemma 48 and Lemma 49 give with $r^* = r^*(x_1, r)$,

$$\begin{aligned} f(x_1 + r^*) \cdot (r - r^*) \approx h(x_1, r) &\approx \frac{1}{|F'(x_1 + r^*)|} \sqrt{f(x_1 + r^*)^2 - f(x_1)^2} \\ &\approx \frac{1}{|F'(x_1)|} \left(\int_{x_1}^{x_1 + r^*} 2f(u)^2 |F'(u)|\, du \right)^{\frac{1}{2}} \\ &\approx \frac{1}{|F'(x_1)|} \left[2f(x_1 + r^*)^2 |F'(x_1)| \cdot r^* \right]^{\frac{1}{2}} \\ &\approx \frac{\sqrt{r^*} f(x_1 + r^*)}{\sqrt{|F'(x_1)|}}, \end{aligned}$$

where we have used Part 2 of Lemma 38 and the fact $r^* = r^*(x_1, r) < r \leq \frac{1}{|F'(x_1)|}$. This implies

$$[|F'(x_1)| (r - r^*)]^2 \approx |F'(x_1)| r^*.$$

Thus

$$[|F'(x_1)| (r - r^*)]^2 + |F'(x_1)| (r - r^*) | \approx |F'(x_1)| r \leq 1.$$

As a result, we have $|F'(x_1)|(r-r^*) \approx |F'(x_1)| r \implies r - r^* \approx r$. This also gives the estimate for h by Lemma 48 since we already have $f(x_1 + r^*) \approx f(x_1)$.

3.5. Area of balls centered at an arbitrary point. In the following proposition we obtain an estimate, similar to (7.9), for areas of balls centered at arbitrary points.

PROPOSITION 51. *Let $P = (x_1, x_2) \in \mathbb{R}^2$ and $r > 0$. Set*

$$B_+(P,r) \equiv \{(y_1, y_2) \in B(P,r) : y_1 > x_1 + r^*\}.$$

If $r \geq \frac{1}{|F'(x_1)|}$ then we recover (7.9)

$$|B(P,r)| \approx \frac{f(x_1+r)}{|F'(x_1+r)|^2} \approx |B_+(P,r)|.$$

On the other hand, if $r \leq \frac{1}{|F'(x_1)|}$ we have

$$|B(P,r)| \approx r^2 f(x_1) \approx |B_+(P,r)|$$

PROOF. Because of symmetry, it is enough to consider $x_1 > 0$ and $y_1 = 0$. So let $P_1 = (x_1, 0)$ with $x_1 > 0$.

Case $r \geq \frac{1}{|F'(x_1)|}$. In this case we will compare the ball $B(P,r)$ to the ball $B(0,R)$ centered at the origin with radius $R = x_1 + r$. First we note that $B(P,r) \subset B(0,R)$ since if $(x,y) \in B(P,r)$, then

$$\text{dist}((0,0),(x,y)) \leq \text{dist}((0,0),P) + \text{dist}(P,(x,y)) < x_1 + r = R.$$

Thus from (7.9) we have

$$|B(P,r)| \leq |B(0,R)| \approx \frac{f(R)}{|F'(R)|^2} = \frac{f(x_1+r)}{|F'(x_1+r)|^2},$$

By parts (2) and (3) of Proposition 47, $h \approx \frac{f(x_1+r)}{|F'(x_1+r)|}$ and $r - r^* \approx \frac{1}{|F'(x_1+r)|}$ when $r \geq \frac{1}{|F'(x_1)|}$, and so we have

$$|B(P,r)| \lesssim h(x_1,r)\ (r - r^*(x_1,r)).$$

Finally, we claim that

$$h(x_1,r)\ (r - r^*(x_1,r)) \lesssim |B(P,r)|.$$

To see this we consider x satisfying $x_1 + r^* \leq x \leq x_1 + \frac{r+r^*}{2}$, where $x_1 + \frac{r+r^*}{2}$ is the midpoint of the interval $[x_1 + r^*, x_1 + r]$ corresponding to the "thick" part of the ball $B(P,r)$. For such x we let $y > 0$ be defined so that $(x,y) \in \partial B(P,r)$. Then using the taxicab path $(x_1,0) \to (x,0) \to (x,y)$, we see that

$$x - x_1 + \frac{y}{f(x)} \geq \text{dist}((x_1,0),(x,y)) = r, \tag{7.14}$$

implies

$$y \geq f(x)(r - x + x_1) \approx f(x_1+r)(r-r^*),$$

where the final approximation follows from $r - r^* \approx \frac{1}{|F'(x_1+r)|}$ and part (2) of Lemma 38 upon using $x_1 + r^* \leq x \leq x_1 + \frac{r+r^*}{2}$. Thus, using parts (2) and (3) of Proposition 47 again, we obtain

$$\begin{aligned} |B(P,r)| &\gtrsim [f(x_1+r)(r-r^*)]\ (r-r^*) \\ &\approx \frac{f(x_1+r)}{|F'(x_1+r)|}(r-r^*) \approx h(x_1,r)\ (r - r^*(x_1,r)), \end{aligned}$$

which proves our claim and concludes the proof that $|B(P,r)| \approx \frac{f(x_1+r)}{|F'(x_1+r)|^2} \approx |B_+(P,r)|$ when $r \geq \frac{1}{|F'(x_1)|}$.

Case $r < \frac{1}{|F'(x_1)|}$. In this case parts (2) and (3) of Proposition 47 show that $h \approx rf(x_1)$ and $r - r^* \approx r$, and part (1) shows that h maximizes the 'height' of the ball. Thus we immediately obtain the upper bound

$$|B(P,r)| \lesssim hr \lesssim f(x_1)\, r^2.$$

To obtain the corresponding lower bound, we use notation as in the first case and note that (7.14) now implies

$$y \geq f(x)(r - x + x_1) \approx f(x_1)\, r, \tag{7.15}$$

where the final approximation follows from part (2) of Proposition 47 and part (2) of Lemma 38 upon using $x_1 + r^* \leq x \leq x_1 + \frac{r+r^*}{2}$. Thus

$$|B(P,r)| \gtrsim [f(x_1)\, r]\ (r - r^*) \approx f(x_1)\, r^2,$$

which concludes the proof that $|B(P,r)| \approx f(x_1)\, r^2 \approx |B_+(P,r)|$ when $r < \frac{1}{|F'(x_1)|}$. □

Using Proposition 47 we obtain a useful corollary for the measure of the "thick" part of a ball. But first we need to establish that $r^*(x_1, r)$ is increasing in r where

$$T(x_1, r) \equiv (x_1 + r^*(x_1, r), h(x_1, r))$$

is the turning point for the geodesic γ_r that passes through $P = (x_1, 0)$ in the upward direction and has vertical slope at the boundary of the ball $B(P,r)$.

LEMMA 52. *Let $x_1 > 0$. Then $r^*(x_1, r') < r^*(x_1, r)$ if $0 < r' < r$.*

PROOF. Let $T(x_1, r) \equiv (x_1 + r^*(x_1, r), h(x_1, r))$ be the turning point for the geodesic γ_r that passes through $P = (x_1, 0)$ and has vertical slope at the boundary of the ball $B(P,r)$. A key property of this geodesic is that it continues beyond the point $T(x_1, r)$ by vertical reflection. Now we claim that this key property implies that when $0 < r' < r$, the geodesic $\gamma_{r'}$ cannot lie below γ_r just to the right of P. Indeed, if it did, then since $B(P, r') \subset B(P,r)$ implies $h(x_1, r') < h(x_1, r)$, the geodesic $\gamma_{r'}$ would turn back and intersect γ_r in the first quadrant, contradicting the fact that geodesics cannot intersect twice in the first quadrant. Thus the geodesic $\gamma_{r'}$ lies above γ_r just to the right of P, and it is now evident that $\gamma_{r'}$ must turn back 'before' γ_r, i.e. that $r^*(x_1, r') < r^*(x_1, r)$. □

COROLLARY 53. *Let $x_1 > 0$. Denote*

$$\begin{aligned} B_+(P,r) &\equiv \{(y_1, y_2) \in B(P,r) : y_1 > x_1 + r^*\}, \\ B_-(P,r) &\equiv \{(y_1, y_2) \in B(P,r) : y_1 \leq x_1 + r^*\}. \end{aligned}$$

Then

$$|B_+(P,r)| \approx |B_-(P,r)| \approx |B(P,r)|.$$

PROOF. **Case** $r < \frac{1}{|F'(x_1)|}$. Recall from Assumption (4) that $\frac{1}{|F'(x_1)|} \leq \frac{1}{\varepsilon} x_1$, so that in this case we have $r < \frac{1}{\varepsilon} x_1$, and hence also that $x_1 - \max\left\{\varepsilon x_1, x_1 - \frac{r}{2}\right\} \approx r$. From Proposition 51 we have

$$|B(P,r)| \approx r^2 f(x_1).$$

From part (2) of Lemma 38, there is a positive constant c such that $f(x) \geq cf(x_1)$ for $\max\left\{\varepsilon x_1, x_1 - \frac{r}{2}\right\} \leq x \leq x_1$. It follows that $B_-(P,r) \supset \left(\max\left\{\varepsilon x_1, x_1 - \frac{r}{2}\right\}, x_1\right) \times \left(-\frac{c}{2}f(x_1)r, \frac{c}{2}f(x_1)r\right)$ since

$$\begin{aligned} d\left((x_1,0),(x,y)\right) &\leq d\left((x_1,0),(x,0)\right) + d\left((x,0),(x,y)\right) \\ &= |x_1 - x| + \frac{|y|}{f(x)} < \frac{r}{2} + \frac{r}{2} = r, \end{aligned}$$

provided $\max\left\{\varepsilon x_1, x_1 - \frac{r}{2}\right\} < x < x_1$ and $-\frac{c}{2}f(x_1)r < y < \frac{c}{2}f(x_1)r$. Thus we have

$$|B_-(P,r)| \geq cr^2 f(x_1).$$

Case $r \geq \frac{1}{|F'(x_1)|}$. The bound $|B_-(P,r)| \leq |B(P,r)| \approx |B_+(P,r)|$ follows from Proposition 51. We now consider two subcases in order to obtain the lower bound $|B_-(P,r)| \gtrsim |B(P,r)|$.

Subcase $r \geq \frac{1}{|F'(x_1)|} \geq r^*$. By (7.11) and part (2) of Lemma 38 we have

$$|F'(x_1+r)| \approx |F'(x_1+r^*)| \text{ and } f(x_1+r) \approx f(x_1+r^*).$$

Then by Proposition 51, followed by the above inequalities, and then another application of part (2) of Lemma 38, we obtain

$$|B(P,r)| \approx \frac{f(x_1+r)}{|F'(x_1+r)|^2} \approx \frac{f(x_1+r^*)}{|F'(x_1+r^*)|^2} \approx \frac{f(x_1)}{|F'(x_1)|^2}.$$

On the other hand, with $r_0 = \frac{1}{|F'(x_1)|}$, we can apply the case already proved above, together with the fact that $m(x_1,r)$ is increasing in r, to obtain that

$$\begin{aligned} |B_-(P,r)| &\geq |\{(y_1,y_2) \in B(P,r_0) : y_1 \leq x_1 + r^*\}| \\ &\geq \left|\{(y_1,y_2) \in B(P,r_0) : y_1 \leq x_1 + (r_0)^*\}\right| \approx \frac{f(x_1)}{|F'(x_1)|^2}. \end{aligned}$$

Subcase $r \geq r^* \geq \frac{1}{|F'(x_1)|}$. Since $B(P,r^*) \subset B_-(P,r)$ we can apply Proposition 51 to $B(P,r^*)$ to obtain

$$|B_-(P,r)| \geq |B(P,r^*)| \approx \frac{f(x_1+r^*)}{|F'(x_1+r^*)|^2}.$$

Now we apply (7.12) and Proposition 51 again to conclude that

$$\frac{f(x_1+r^*)}{|F'(x_1+r^*)|^2} \approx \frac{f(x_1+r)}{|F'(x_1+r)|^2} \approx |B(P,r)|.$$

□

4. Higher dimensional geometries

First we consider the 3-dimensional case.

4.1. Geodesics and metric balls. Let $\gamma(t) = (x_1(t), x_2(t), x_3(t))$ be a path. Then the arc length element is given by

$$ds = \sqrt{[x_1'(t)]^2 + [x_2'(t)]^2 + \frac{1}{[f(x_1)]^2}[x_3'(t)]^2 dt}.$$

Thus we can factor the associated control space by

$$\left(\mathbb{R}^3, \begin{bmatrix} 1 & 0 & 0 \\ 0 & 1 & 0 \\ 0 & 0 & [f(x_1)]^{-2} \end{bmatrix}\right) = \left(\mathbb{R}^2_{x_1,x_3}, \begin{bmatrix} 1 & 0 \\ 0 & [f(x_1)]^{-2} \end{bmatrix}\right) \times \mathbb{R}_{x_2}.$$

We begin with a lemma regarding paths in product spaces.

LEMMA 54. *Let (M_1, g^{M_1}) and (M_2, g^{M_2}) be two Riemannian manifolds. Consider the Cartesian product $M_1 \times M_2$ whose Riemannian metric is defined by*

$$g_{(p,q)}((u_1,u_2),(v_1,v_2)) = g_p^{M_1}(u_1,v_1) + g_q^{M_2}(u_2,v_2).$$

Here we have

$$(p,q) \in M_1 \times M_2 \text{ and } (u_1,u_2), (v_1,v_2) \in T_p(M_1) \oplus T_p(M_2) \approx T_{(p,q)}(M_1 \times M_2).$$

Given any C^1 path $\gamma : [a,b] \to M_1 \times M_2$, we can write it in the form $(\gamma_1(t), \gamma_2(t))$, where $\gamma_1 : [a,b] \to M_1$ and $\gamma_2 : [a,b] \to M_2$ are C^1 paths on M_1 and M_2, respectively. Then we have

$$\|\gamma\| \geq \sqrt{\|\gamma_1\|^2 + \|\gamma_2\|^2}$$

where $\|\gamma\|$, $\|\gamma_1\|$ and $\|\gamma_2\|$ represent the arc length of each path. In addition, equality occurs if and only if

$$\frac{\|\gamma_1'(t)\|_{g^{M_1}}}{\|\gamma_1\|} = \frac{\|\gamma_2'(t)\|_{g^{M_2}}}{\|\gamma_2\|}, \qquad a \leq t \leq b. \tag{7.16}$$

PROOF. For simplicity we omit the subscripts of the norms $\|\gamma_1'(t)\|_{g^{M_1}}$ and $\|\gamma_2'(t)\|_{g^{M_2}}$ so that $\|\gamma_j\| = \int_a^b \sqrt{\|\gamma_j'(t)\|^2} dt$. Using that

$$\frac{\|\gamma_1\|}{\sqrt{\|\gamma_1\|^2 + \|\gamma_2\|^2}} \|\gamma_1'(t)\| + \frac{\|\gamma_2\|}{\sqrt{\|\gamma_1\|^2 + \|\gamma_2\|^2}} \|\gamma_2'(t)\| \leq \sqrt{\|\gamma_1'(t)\|^2 + \|\gamma_2'(t)\|^2},$$

with equality if and only if

$$\begin{pmatrix} \|\gamma_1'(t)\| \\ \|\gamma_2'(t)\| \end{pmatrix} \text{ is parallel to } \begin{pmatrix} \|\gamma_1\| \\ \|\gamma_2\| \end{pmatrix},$$

we obtain that

$$\begin{aligned} \|\gamma\| &= \int_a^b \sqrt{\|\gamma_1'(t)\|^2 + \|\gamma_2'(t)\|^2} dt \\ &\geq \int_a^b \left(\frac{\|\gamma_1\|}{\sqrt{\|\gamma_1\|^2 + \|\gamma_2\|^2}} \|\gamma_1'(t)\| + \frac{\|\gamma_2\|}{\sqrt{\|\gamma_1\|^2 + \|\gamma_2\|^2}} \|\gamma_2'(t)\| \right) dt \\ &= \frac{\|\gamma_1\|^2}{\sqrt{\|\gamma_1\|^2 + \|\gamma_2\|^2}} + \frac{\|\gamma_2\|^2}{\sqrt{\|\gamma_1\|^2 + \|\gamma_2\|^2}} = \sqrt{\|\gamma_1\|^2 + \|\gamma_2\|^2}, \end{aligned}$$

with equality if and only if (7.16) holds. □

COROLLARY 55. *A C^1 path $\gamma = (\gamma_1, \gamma_2)$ is a geodesic of $M_1 \times M_2$ if and only if*

(1) *γ_1 is a geodesic of M_1,*
(2) *γ_2 is a geodesic of M_2,*
(3) *and the speeds of γ_1 and γ_2 match, i.e. the identity $\frac{\|\gamma_1'(t)\|_{g^{M_1}}}{\|\gamma_1\|} = \frac{\|\gamma_2'(t)\|_{g^{M_2}}}{\|\gamma_2\|}$ holds for all t.*

COROLLARY 56. *The distance between two points $(p_1, q_1), (p_2, q_2) \in M_1 \times M_2$ is given by*

$$d_g((p_1, q_1), (p_2, q_2)) = \sqrt{\left[d_{g^{M_1}}(p_1, p_2)\right]^2 + \left[d_{g^{M_2}}(q_1, q_2)\right]^2}.$$

Thus we can write a typical geodesic in the form

$$\begin{cases} x_2 = C_2 \pm k \int_0^{x_1} \frac{\lambda}{\sqrt{\lambda^2 - [f(u)]^2}} du \\ x_3 = C_3 \pm \int_0^{x_1} \frac{[f(u)]^2}{\sqrt{\lambda^2 - [f(u)]^2}} du \end{cases},$$

and a metric ball centered at $y = (y_1, y_2, y_3)$ with radius $r > 0$ is given by

$$B(y,r) \equiv \left\{ (x_1, x_2, x_3) : (x_1, x_3) \in B_{2D}\left((y_1, y_3), \sqrt{r^2 - |x_2 - y_2|^2} \right) \right\},$$

where $B_{2D}(a, s)$ denotes the 2-dimensional control ball centered at a in the plane with radius s that was associated with f above.

4.2. Volumes of n-dimensional balls. Recall that the Lebesgue measure of the *two* dimensional ball $B_{2D}(x, r)$ satisfies

$$|B_{2D}(x,r)| \approx \begin{cases} r^2 f(x_1) & \text{if} \quad r \leq \frac{1}{|F'(x_1)|} \\ \frac{f(x_1+r)}{|F'(x_1+r)|^2} & \text{if} \quad r \geq \frac{1}{|F'(x_1)|} \end{cases}.$$

Recall also that in the two dimensional case, we had

$$|B_{2D}(x, d(x,y))| \approx h_{x,y} \widehat{d}(x,y) \approx h_{x,y} \min\left\{ d(x,y), \frac{1}{|F'(x_1 + d(x,y))|} \right\}.$$

In the three dimensional case, the quantities $h_{x,y}$ and $\widehat{d}(x,y)$ remain formally the same and as was done above, we can write a typical geodesic in the form

$$\begin{cases} x_2 = C_2 \pm k \int_0^{x_1} \frac{\lambda}{\sqrt{\lambda^2 - [f(u)]^2}} du \\ x_3 = C_3 \pm \int_0^{x_1} \frac{[f(u)]^2}{\sqrt{\lambda^2 - [f(u)]^2}} du \end{cases},$$

so that a metric ball centered at $y = (y_1, y_2, y_3)$ with radius $r > 0$ is given by

$$B(y,r) \equiv \left\{ (x_1, x_2, x_3) : (x_1, x_3) \in B_{2D}\left((y_1, y_3), \sqrt{r^2 - |x_2 - y_2|^2} \right) \right\},$$

where $B_{2D}(a, s)$ denotes the 2-dimensional control ball centered at a in the plane parallel to the x_1, x_3-plane with radius s that was associated with f above.

In dimension $n \geq 4$, the same arguments show that a typical geodesic has the form

$$\begin{cases} \mathbf{x}_2 = \mathbf{C}_2 \pm \mathbf{k} \int_0^{x_1} \frac{\lambda}{\sqrt{\lambda^2 - [f(u)]^2}} du \\ x_3 = C_3 \pm \int_0^{x_1} \frac{[f(u)]^2}{\sqrt{\lambda^2 - [f(u)]^2}} du \end{cases},$$

where $\mathbf{x}_2, \mathbf{C}_2, \mathbf{k} \in \mathbb{R}^{n-2}$ are now $(n-2)$-dimensional vectors, so that a metric ball centered at

$$y = (y_1, \mathbf{y}_2, y_3) \in \mathbb{R} \times \mathbb{R}^{n-2} \times \mathbb{R} = \mathbb{R}^n,$$

with radius $r > 0$ is given by

$$B(y,r) \equiv \left\{ (x_1, \mathbf{x}_2, x_3) : (x_1, x_3) \in B_{2D}\left((y_1, y_3), \sqrt{r^2 - |\mathbf{x}_2 - \mathbf{y}_2|^2} \right) \right\},$$

where $B_{2D}(a,s)$ denotes the 2-dimensional control ball centered at a in the plane parallel to the x_1, x_3-plane with radius s that was associated with f above.

LEMMA 57. *The Lebesgue measure of the* three *dimensional ball* $B_{3D}(x,r)$ *satisfies*

$$|B_{3D}(x,r)| \approx \begin{cases} r^3 f(x_1) & \text{if} \quad r \leq \frac{2}{|F'(x_1)|} \\ \frac{f(x_1+r)}{|F'(x_1+r)|^3} \sqrt{r\,|F'(x_1+r)|} & \text{if} \quad r \geq \frac{2}{|F'(x_1)|} \end{cases},$$

and that of the n-dimensional ball $B_{nD}(x,r)$ *satisfies*

$$|B_{nD}(x,r)| \approx \begin{cases} r^n f(x_1) & \text{if} \quad r \leq \frac{2}{|F'(x_1)|} \\ \frac{f(x_1+r)}{|F'(x_1+r)|^n} \left(r\,|F'(x_1+r)|\right)^{\frac{n}{2}-1} & \text{if} \quad r \geq \frac{2}{|F'(x_1)|} \end{cases},$$

PROOF. We estimate the measure $|B(x,r)|$ of an n-dimensional ball $B(x,r) = B_{nD}(x,r)$, where $x = (x_1, \mathbf{x}_2, x_3) \in \mathbb{R} \times \mathbb{R}^{n-2} \times \mathbb{R} = \mathbb{R}^n$, and where we use boldface font for $\mathbf{x}_2$ to emphasize that it belongs to $\mathbb{R}^{n-2}$ as opposed to $\mathbb{R}$. We consider two cases, where we may assume by symmetry that $\mathbf{x}_2 = \mathbf{0}$ and $x_3 = 0$.

Case $r < \frac{2}{|F'(x_1)|}$: In this case we have $\sqrt{r^2 - |\mathbf{y}_2|^2} \leq r < \frac{2}{|F'(x_1)|}$ and

$$\left| B_{2D}\left((x_1, \mathbf{0}, 0), \sqrt{r^2 - |\mathbf{y}_2|^2} \right) \right| \approx (r^2 - |\mathbf{y}_2|^2) f(x_1),$$

where for $r < \frac{1}{|F'(x_1)|}$ we appeal to the second assertion in Proposition 51, while for $\frac{1}{|F'(x_1)|} \leq r < \frac{2}{|F'(x_1)|}$ we appeal to the first assertion in Proposition 51 and use the estimates for f and $|F'|$ in (2) of Lemma 38. With $\mathrm{A}(a,b) \equiv \{\mathbf{y}_2 \in \mathbb{R}^{n-2} : a \leq |\mathbf{y}_2| \leq b\}$ denoting the annulus centered at the origin in $\mathbb{R}^{n-2}$ with radii $a < b$, the above gives

$$\begin{aligned} |B(x,r)| = \int_{\mathrm{A}(0,r)} \left| B_{2D}\left((x_1, \mathbf{0}, 0), \sqrt{r^2 - |\mathbf{y}_2|^2} \right) \right| d\mathbf{y}_2 &\approx \int_{\mathrm{A}(0,r)} (r^2 - |\mathbf{y}_2|^2) f(x_1) d\mathbf{y}_2 \\ &\approx r^n f(x_1). \end{aligned}$$

Case $r \geq \frac{2}{|F'(x_1)|}$: Again we have

$$|B(x,r)| = \int_{B(0,r)} \left| B_{2D}\left((x_1, \mathbf{0}, 0), \sqrt{r^2 - |\mathbf{y}_2|^2} \right) \right| d\mathbf{y}_2 \,.$$

Since the measure of the ball $B_{2D}((x_1,\mathbf{0},0),R)$ is nondecreasing as a function of the radius R, we have

$$|B(x,r)| \approx \int_{B(0,\frac{r}{2})} \left| B_{2D}\left((x_1,\mathbf{0},0),\sqrt{r^2-|\mathbf{y}_2|^2}\right)\right| d\mathbf{y}_2$$

$$= \int_{B(0,\frac{r}{2})} \frac{f\left(x_1+\sqrt{r^2-|\mathbf{y}_2|^2}\right)}{\left|F'\left(x_1+\sqrt{r^2-|\mathbf{y}_2|^2}\right)\right|^2} d\mathbf{y}_2 .$$

Using polar coordinates and our assumptions on F' we continue with

$$|B(x,r)| \approx \int_0^{\frac{r}{2}} \frac{f\left(x_1+\sqrt{r^2-R^2}\right)}{\left|F'\left(x_1+\sqrt{r^2-R^2}\right)\right|^2} R^{n-3}dR$$

$$\approx \frac{1}{|F'(x_1+r)|^2}\int_0^{\frac{r}{2}} f\left(x_1+\sqrt{r^2-R^2}\right) R^{n-3}dR.$$

Now use the change of variable $w=r-\sqrt{r^2-R^2}\in\left(0,\frac{2-\sqrt{3}}{2}r\right)$ to write the last integral as

$$\int_0^{\frac{r}{2}} f\left(x_1+\sqrt{r^2-R^2}\right) R^{n-3}dR \approx \int_0^{\frac{2-\sqrt{3}}{2}r} f(x_1+r-w)(rw)^{\frac{n}{2}-2}rdw,$$

so that we obtain

$$|B(x,r)| \approx \frac{r^{\frac{n}{2}-1}}{|F'(x_1+r)|^2}\int_0^{\frac{2-\sqrt{3}}{2}r} f(x_1+r-w)\, w^{\frac{n}{2}-2}dw.$$

Now we observe that the upper limit of the integral above satisfies

$$\frac{2-\sqrt{3}}{2}r \gtrsim \frac{1}{|F'(x_1+r)|}.$$

Indeed, if $r<x_1$, then our assumption on r gives $\frac{2-\sqrt{3}}{2}r \gtrsim \frac{1}{|F'(x_1)|}\approx\frac{1}{|F'(x_1+r)|}$, while if $r\geq x_1$, then our assumption on F' gives $\frac{2-\sqrt{3}}{2}r\gtrsim x_1+r\gtrsim\frac{1}{|F'(x_1+r)|}$. Lemma 57 now follows immediately from this equivalence:

For $\beta>-1$, $0<\varepsilon<1$ and $\frac{\varepsilon}{|F'(z_1)|}<r<z_1$, we have

$$\int_0^r f(z_1-w)\, w^\beta dw \approx \frac{f(z_1)}{|F'(z_1)|^{\beta+1}}. \tag{7.17}$$

To see (7.17), we note that on the one hand,

$$\int_0^r f(z_1-w)\, w^\beta dw \geq \int_0^{\frac{\varepsilon}{|F'(z_1)|}} f(z_1-w)\, w^\beta dw \approx \frac{f(z_1)}{|F'(z_1)|^{\beta+1}}.$$

On the other hand,

$$\begin{aligned}\ln \frac{f(z_1 - w)}{f(z_1)} &= \int_{z_1 - w}^{z_1} F'(t)\, dt \le -\left|F'(z_1)\right| w \\ &\Longrightarrow f(z_1 - w) \le f(z_1)\, e^{-\left|F'(z_1)\right| w},\end{aligned}$$

which gives

$$\int_0^{z_1} f(z_1 - w)\, w^{\beta} dw \lesssim \frac{f(z_1)}{\left|F'(z_1)\right|^{\beta+1}}.$$

$\square$

CHAPTER 8

Orlicz Norm Sobolev Inequalities

In this second chapter of Part 3, we prove Orlicz Sobolev and proportional vanishing L^1-Sobolev inequalities for infinitely degenerate geometries. The key to these inequalities is a subrepresentation formula for a Lipschitz function w in terms of its control gradient that vanishes at the 'end' of a ball. The kernel $K(x,y)$ of this subrepresentation in the infinitely degenerate setting is in general much smaller that the familiar $\frac{\text{distance}}{\text{volume}} = \frac{d(x,y)}{|B(x,d(x,y))|}$ kernel that arises in the finite type case - see Remark 68 below for more on this. With this we then establish Orlicz Sobolev bump inequalities and the more familiar $1-1$ Poincaré inequality. This latter inequality is then extended to a *proportional vanishing* subrepresentation inequality for w, where w vanishes on a subset of measure proportional to that of the ball, and then to a proportional vanishing L^1-Sobolev inequality.

1. Subrepresentation inequalities

We first consider the two dimensional case, and then generalize to higher dimensions in the subsequent subsection.

1.1. The 2-dimensional case. We will obtain a subrepresentation formula for the degenerate geometry by applying the method of Lemma 79 in [**SaWh4**]. For simplicity, we will only consider x with $x_1 > 0$; since our metric is symmetric about the y axis it suffices to consider this case. For the general case, all objects defined on the right half plane must be defined on the left half plane by reflection about the y-axis.

Consider a sequence of control balls $\{B(x,r_k)\}_{k=1}^{\infty}$ centered at x with radii $r_k \searrow 0$ such that $r_0 = r$ and

$$|B(x,r_k) \setminus B(x,r_{k+1})| \approx |B(x,r_{k+1})|, \qquad k \geq 1,$$

so that $B(x,r_k)$ is divided into two parts having comparable area. We may in fact assume that

$$r_{k+1} = \begin{cases} r^*(x_1,r_k) & \text{if } r_k \geq \frac{1}{|F'(x_1)|} \\ \frac{1}{2} r_k & \text{if } r_k < \frac{1}{|F'(x_1)|} \end{cases} \tag{8.1}$$

where r^* is defined in Proposition 47. Indeed, if $r_k \geq \frac{1}{|F'(x_1)|}$, then by (1) in Proposition 47 we have that

$$r_k - r_{k+1} \approx \frac{1}{|F'(x_1 + r_k)|}$$

and then by (2) in Lemma 38 it follows that $f(x_1 + r_k) \approx f(x_1 + r_{k+1})$ and $|F'(x_1 + r_k)| \approx |F'(x_1 + r_{k+1})|$, so by Corollary 53 and (1) in Proposition 47 it follows that

$$\begin{aligned}
|B(x,r_k)| &\approx \left|\left\{B(x,r_k)\bigcap y_1 > x_1 + r_{k+1}\right\}\right| \approx (r_k - r_{k+1})\, h(x_1, x_1 + r_k) \\
&\approx \frac{1}{|F'(x_1+r_k)|}\frac{f(x_1+r_k)}{|F'(x_1+r_k)|} \approx \frac{1}{|F'(x_1+r_{k+1})|}\frac{f(x_1+r_{k+1})}{|F'(x_1+r_{k+1})|} \\
&\approx (r_{k+1} - r_{k+2})\, h(x_1, x_1 + r_{k+1}) \\
&\approx \left|\left\{B(x,r_{k+1})\bigcap y_1 > x_1 + r_{k+2}\right\}\right| \approx |B(x,r_{k+1})|\,.
\end{aligned}$$

On the other hand, if $r_k \leq \frac{1}{|F'(x_1)|}$ then by (2) in Proposition 47 $r_k - r_{k+1} \approx r_k$ and $h(x_1, x_1 + r_k) \approx r_k f(x_1)$, hence by Corollary 53

$$|B(x,r_k)| \approx (r_k - r_{k+1})\, h(x_1, x_1 + r_{k+1}) \approx r_k^2 f(x_1) \approx r_{k+1}^2 f(x_1) \approx |B(x, r_{k+1})|\,.$$

As a consequence we also have that

$$\begin{aligned}
(r_k - r_{k+1})\, h(x_1, x_1 + r_k) &\approx (r_{k+1} - r_{k+2})\, h(x_1, x_1 + r_{k+1}) \\
&\lesssim (r_{k+1} - r_{k+2})\, h(x_1, x_1 + r_k)
\end{aligned}$$

so $r_k - r_{k+1} \leq C(r_{k+1} - r_{k+2}) \leq C r_{k+1}$, which yields

$$\frac{1}{C+1} r_k \leq r_{k+1}. \tag{8.2}$$

Now for $x_1, t > 0$ define

$$h^*(x_1, t) = \int_{x_1}^{x_1+t} \frac{f^2(u)}{\sqrt{f^2(x_1+t) - f^2(u)}}\, du, \tag{8.3}$$

so that $h^*(x_1, t)$ describes the 'height' above x_2 at which the geodesic through $x = (x_1, x_2)$ curls back toward the y-axis at the point $(x_1 + t, x_2 + h^*(x_1, t))$. Thus the graph of $y = h^*(x_1, t)$ is the curve separating the analogues of Region 1 and Region 2 relative to the ball $B(x, r)$. See Figure 3.

Then in the case $r_k \geq \frac{1}{|F'(x_1)|}$, we have $h^*(x_1, r_{k+1}) = h(x_1, r_k)$, $k \geq 0$, where $h(x_1, r_k)$ is the height of $B(x, r_k)$ as defined in Proposition 47. In the opposite case $r_k < \frac{1}{|F'(x_1)|}$, we have $r_{k+1} = \frac{1}{2} r_k$ instead, and we will estimate differently.

For $k \geq 0$ define

$$\begin{aligned}
&E(x, r_k) \\
&\equiv \begin{cases} \{y : x_1 + r_{k+1} \leq y_1 < x_1 + r_k,\ |y_2| < h^*(x_1, y_1 - x_1)\}, & \text{if } r_k \geq \frac{1}{|F'(x_1)|} \\ \{y : x_1 + r_{k+1} \leq y_1 < x_1 + r_k,\ |y_2| < h^*(x_1, r_k^*) = h(x_1, r_k)\}, & \text{if } r_k < \frac{1}{|F'(x_1)|} \end{cases}
\end{aligned}$$

where we have written $r_k^* = r^*(x_1, r_k)$ for convenience. We claim that

$$|E(x, r_k)| \approx \left|E(x, r_k) \bigcap B(x, r_k)\right| \approx |B(x, r_k)| \text{ for all } k \geq 1. \tag{8.4}$$

Indeed, in the first case $r_k \geq \frac{1}{|F'(x_1)|}$, the second set of inequalities follows immediately by Corollary 53, and since $E(x, r_k) \subset B(x, r_{k-1})$ we have that

$$\begin{aligned}
\left|E(x, r_k) \bigcap B(x, r_k)\right| &\leq |E(x, r_k)| \leq |B(x, r_{k-1})| \\
&\lesssim |B(x, r_k)| \lesssim \left|E(x, r_k) \bigcap B(x, r_k)\right|,
\end{aligned}$$

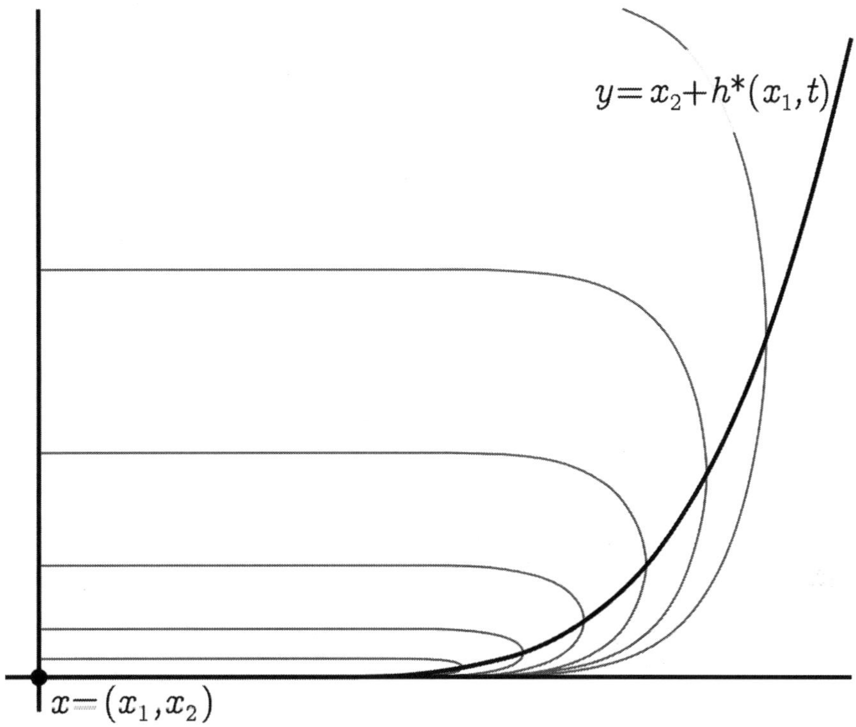

FIGURE 3. The graph of h^* for $x_1 > 0$ small.

which establishes the first set of inequalities in (8.4). In the second case $r_k < \frac{1}{|F'(x_1)|}$, we have

$$|E(x, r_k)| = \frac{1}{2} r_k h^*(x_1, r_k^*) \approx (r_k - r_k^*) h(x_1, r_k^*) \approx |B(x, r_k)|,$$

and from (7.15) with $(x, y) \in \partial B(x, r_k)$, we have

$$y \geq f(x)(r_k - x + x_1) \approx f(x_1) r_k,$$

for *all* $x \in [x_1, x_1 + r]$ since we are in the case $r_k < \frac{1}{|F'(x_1)|}$. It follows that

$$E(x, r_k) \cap B(x, r_k) \supset \left[x_1 + \frac{r_k}{2}, x_1 + \frac{3r_k}{4}\right] \times [-cf(x_1) r_k, cf(x_1) r_k]$$

and hence that

$$|E(x, r_k) \cap B(x, r_k)| \geq \frac{1}{2} c r_k f(x_1) r_k \approx |B(x, r_k)| \geq |E(x, r_k) \cap B(x, r_k)|.$$

This completes the proof of (8.4).

Now define $\Gamma(x, r)$ to be the set

$$\Gamma(x, r) = \bigcup_{k=1}^{\infty} \operatorname{co}[E(x, r_k) \cup E(x, r_{k+1})],$$

where co$E$ denotes the convex hull of the set E. Set

$$\mathbb{E}_{x,r_1} w \equiv \frac{1}{|E(x,r_1)|} \int\int_{E(x,r_1)} w.$$

LEMMA 58. *With $\Gamma(x,r)$, $E(x,r_1)$ and $\mathbb{E}_{x,r_1}$ as above, and in particular with $r_0 = r$ and r_1 given by (8.1), we have the subrepresentation formula*

$$|w(x) - \mathbb{E}_{x,r_1} w| \leq C \int_{\Gamma(x,r)} |\nabla_A w(y)| \frac{\widehat{d}(x,y)}{|B(x,d(x,y))|} dy, \tag{8.5}$$

where ∇_A is as in (1.4) and

$$\widehat{d}(x,y) \equiv \min\left\{ d(x,y), \frac{1}{|F'(x_1 + d(x,y))|} \right\}. \tag{8.6}$$

Note that when $f(r) = r^N$ is finite type, then $\widehat{d}(x,y) \approx d(x,y)$.

PROOF. Recall the sequence $\{r_k\}_{k=1}^{\infty}$ of decreasing radii above. Then since w is *a priori* Lipschitz continuous, we can write,

$$\begin{aligned} w(x) - &\mathbb{E}_{x,r_1} w \\ &= \lim_{k\to\infty} \frac{1}{|E(x,r_k)|} \int_{E(x,r_k)} w(y)\,dy - \frac{1}{|E(x,r_1)|} \int_{E(x,r_1)} w \\ &= \sum_{k=1}^{\infty} \left\{ \frac{1}{|E(x,r_{k+1})|} \int_{E(x,r_{k+1})} w(y)\,dy - \frac{1}{|E(x,r_k)|} \int_{E(x,r_k)} w(z)\,dz \right\}, \end{aligned}$$

and so we have

$$\begin{aligned} |w(x) &- \mathbb{E}_{x,r_1} w| \\ &\lesssim \sum_{k=1}^{\infty} \frac{1}{|B(x,r_k)|^2} \int_{E(x,r_{k+1})\times E(x,r_k)} |w(y) - w(z)|\,dydz \\ &\lesssim \sum_{k=1}^{\infty} \frac{1}{|B(x,r_k)|^2} \int_{E(x,r_{k+1})\times E(x,r_k)} \\ &\qquad \times \{ |w(y_1,y_2) - w(z_1,y_2)| + |w(z_1,y_2) - w(z_1,z_2)| \}\,dydz \\ &\lesssim \sum_{k=1}^{\infty} \frac{1}{|B(x,r_k)|^2} \int_{E(x,r_{k+1})\times E(x,r_k)} \int_{y_1}^{z_1} |w_x(s,y_2)|\,ds dy_1 dy_2 dz_1 dz_2 \\ &\quad + \sum_{k=1}^{\infty} \frac{1}{|B(x,r_k)|^2} \int_{E(x,r_{k+1})\times E(x,r_k)} \int_{y_2}^{z_2} |w_y(z_1,t)|\,dt dy_1 dy_2 dz_1 dz_2\ , \end{aligned}$$

which, with $H_k(x) \equiv E(x,r_{k+1}) \bigcup E(x,r_k)$, is dominated by

$$\begin{aligned} &\sum_{k=1}^{\infty} \frac{1}{|B(x,r_k)|^2} \left(\int_{H_k(x)} |\nabla_A w(s,y_2)|\,ds dy_2 \right) \{r_k - r_{k+1}\} \int_{H_k(x)} dz_1 dz_2 \\ &+ \sum_{k=1}^{\infty} \frac{1}{|B(x,r_k)|^2} \left(\int_{H_k(x)} |\nabla_A w(z_1,t)|\,dz_1 dt \right) \frac{h_k}{f(x_1 + r_{k+1})} \int_{H_k(x)} dy_1 dy_2\ , \end{aligned}$$

where for the last term we used that

$$\begin{aligned}|w_y(z_1,t)| &= \frac{f(z_1)}{f(z_1)}|w_y(z_1,t)| \leq \frac{1}{f(z_1)}|\nabla_A w(z_1,t)| \\ &\leq \frac{1}{f(x_1+r_{k+1})}|\nabla_A w(z_1,t)| \quad \forall (z_1,z_2)\in E(x,r_k).\end{aligned}$$

Next, recall from Lemma 48 that $h_k \approx (r_k - r_{k+1})\cdot f(x_1+r_{k+1})$ by our choice of r_{k+1} in (8.1). Moreover, by the estimates above we have that $|H_k(x)| \approx |B(x,r_k)|$, and

$$\begin{aligned}|w(x) - \mathbb{E}_{x,r_1}w| &\lesssim \sum_{k=1}^{\infty}\frac{r_k - r_{k+1}}{|B(x,r_k)|}\left(\int_{H_k(x)}|\nabla_A w(s,y_2)|\,dsdy_2\right) \\ &\lesssim \int_{\Gamma(x,r)}|\nabla_A w(y)|\left(\sum_{k=1}^{\infty}\frac{r_k-r_{k+1}}{|B(x,r_k)|}\mathbf{1}_{E(x,r_k)}(y)\right)dy. \qquad (8.7)\end{aligned}$$

To make further estimates we need to consider two regions separately, namely;

case 1 $d(x,y) \geq \frac{1}{|F'(x_1)|}$. In this case we have

$$r_k > d(x,y) \geq \frac{1}{|F'(x_1)|},$$

which implies by Proposition 47 and (7.12)

$$r_k - r_{k+1} \approx \frac{1}{|F'(x_1+r_k)|} \approx \frac{1}{|F'(x_1+r_{k+2})|} < \frac{1}{|F'(x_1+d(x,y))|}.$$

Therefore, we are left with

$$\begin{aligned}&|w(x) - \mathbb{E}_{x,r_1}w| \qquad (8.8)\\ &\lesssim \int_{\Gamma(x,r)}|\nabla_A w(y)|\frac{1}{|F'(x_1+d(x,y))|}\sum_{k:r_{k+1}<d(x,y)<r_k}\frac{1}{|B(x,r_k)|}dy \\ &\approx \int_{\Gamma(x,r)}|\nabla_A w(y)|\frac{1}{|F'(x_1+d(x,y))|}\frac{1}{|B(x,d(x,y))|}dy.\end{aligned}$$

case 2 $d(x,y) < \frac{1}{|F'(x_1)|}$. We can write

$$\sum_{k:r_{k+1}<d(x,y)<r_k}\frac{r_k-r_{k+1}}{|B(x,r_k)|} \leq \sum_{k:r_{k+1}<d(x,y)<r_k}\frac{r_k}{|B(x,r_k)|} \lesssim \frac{d(x,y)}{|B(x,d(x,y))|},$$

which gives

$$|w(x) - \mathbb{E}_{x,r_1}w| \lesssim \int_{\Gamma(x,r)}|\nabla_A w(y)|\frac{d(x,y)}{|B(x,d(x,y))|}. \qquad (8.9)$$

To finish the proof we need to dominate the right hand sides of (8.8) and (8.9) with

$$\int_{\Gamma(x,r)}|\nabla_A w(y)|\frac{\widehat{d}(x,y)}{|B(x,d(x,y))|} \qquad (8.10)$$

where $\widehat{d}(x,y) = \min\left\{d(x,y), \frac{1}{|F'(x_1+d(x,y))|}\right\}$ in cases 1 and 2 respectively.

Suppose first that $d(x,y) \geq \frac{1}{|F'(x_1+d(x,y))|}$. Since $|F'(x_1)|$ is a decreasing function of x_1 we have $d(x,y) \geq \frac{1}{|F'(x_1)|}$ and therefore we are in **case 1** and (8.10) then follows from $\frac{1}{|F'(x_1+d(x,y))|} = \widehat{d}(x,y)$.

If the reverse inequality holds, namely $d(x,y) < \frac{1}{|F'(x_1+d(x,y))|}$, we have to consider two subcases. First, if $d(x,y) \leq \frac{1}{|F'(x_1)|}$, then we are in **case 2** and (8.10) then follows from $d(x,y) = \widehat{d}(x,y)$. Finally, if

$$\frac{1}{|F'(x_1)|} \leq d(x,y) < \frac{1}{|F'(x_1+d(x,y))|},$$

we are back in **case 1,** but by Proposition 47 we have

$$\frac{1}{|F'(x_1+d(x,y))|} \approx d(x,y) - d(x,y)^* < d(x,y),$$

and again (8.10) holds since $d(x,y) = \widehat{d}(x,y)$. $\square$

As a simple corollary we obtain a connection between $\widehat{d}(x,y)$ and the 'width' of the thickest part of a ball of radius $d(x,y)$, namely $d(x,y) - d^*(x,y)$, where if $r = d(x,y)$ and r^* is as defined at the beginning of Subsection 3.4 of Chapter 7, then we define $d^*(x,y)$ by

$$d^*(x,y) \equiv r^*. \tag{8.11}$$

Note that if x and r are fixed, then for every $y \in \partial B(x,r)$ we have $d(x,y) - d^*(x,y) = r - r^*$.

COROLLARY 59. *Let $d(x,y) > 0$ be the distance between any two points $x, y \in \Omega$ and let $d^*(x,y)$ be as in (8.11), and $\widehat{d}(x,y)$ be as defined in (8.6) of Lemma 58. Then*

$$\widehat{d}(x,y) \approx d(x,y) - d^*(x,y)$$

PROOF. As before, we consider two cases

case 1 $d(x,y) \geq \frac{1}{|F'(x_1)|}$. In this case we have from Proposition 47

$$d(x,y) - d^*(x,y) \approx \frac{1}{|F'(x_1+d(x,y))|}.$$

If $d(x,y) \geq \frac{1}{|F'(x_1+d(x,y))|}$, then $\widehat{d}(x,y) = \frac{1}{|F'(x_1+d(x,y))|}$ and the claim is proved. If, on the other hand,

$$d(x,y) \leq \frac{1}{|F'(x_1+d(x,y))|},$$

then $\widehat{d}(x,y) = d(x,y)$ and

$$d(x,y) > d(x,y) - d^*(x,y) \approx \frac{1}{|F'(x_1+d(x,y))|} \geq d(x,y),$$

and the claim follows.

case 2 $d(x,y) < \frac{1}{|F'(x_1)|}$. From Proposition 47 we have in this case

$$d(x,y) - d^*(x,y) \approx d(x,y).$$

From the monotonicity of the function $F'(x)$ we have

$$d(x,y) < \frac{1}{|F'(x_1)|} \leq \frac{1}{|F'(x_1+d(x,y))|},$$

and therefore $\widehat{d}(x,y) = d(x,y) \approx d(x,y) - d^*(x,y)$.

□

1.2. The higher dimensional case. The subrepresentation inequality here is similar to Lemma 58 in two dimensions, with the main differences being in the definition of the cusp-like region $\Gamma(x,r)$ in higher dimensions. On the one hand, the shape of the higher dimensional balls dictates the rough form of $\Gamma(x,r)$, but we will also need to redefine the sequence of radii $\{r_k\}_{k=1}^{\infty}$ used in the definition of $\Gamma(x,r)$. We begin by addressing the higher dimensional form, and later will turn to the new sequences $\{r_k\}_{k=1}^{\infty}$.

Recall that we denote points $x \in \mathbb{R}^n$ as

$$x = (x_1, \mathbf{x}_2, x_3) \in \mathbb{R} \times \mathbb{R}^{n-2} \times \mathbb{R}.$$

Let $|B(x,d(x,y))|$ denote the n-dimensional Lebesgue measure of $B(x,d(x,y))$ where $d(x,y)$ is now the n-dimensional control distance. We define the cusp-like region $\Gamma(x,r)$ and the 'ends' $E(x,r_k)$ of the balls $B(x,r_k)$ by

$$\begin{aligned} \Gamma(x,r) &= \bigcup_{k=1}^{\infty} E(x,r_k); \\ E(x,r_k) &\equiv \left\{ y=(y_1,\mathbf{y}_2,y_3) : \begin{array}{c} r_{k+1} \le y_1 - x_1 < r_k \\ |\mathbf{y}_2 - \mathbf{x}_2| < \sqrt{r_k^2 - (y_1-x_1)^2} \\ |y_3 - x_3| < h^*(x_1, y_1 - x_1) \end{array} \right\}, \end{aligned} \tag{8.12}$$

where we recall

$$r_{k+1} = \begin{cases} r^*(x_1, r_k) & \text{if } r_k \ge \frac{1}{|F'(x_1)|} \\ \frac{1}{2} r_k & \text{if } r_k < \frac{1}{|F'(x_1)|} \end{cases},$$

and where r^* is defined in Definition 46 right before Proposition 47. We also define the modified 'end' $\widetilde{E}$ by

$$\widetilde{E}(x,r_k) \equiv \left\{ y=(y_1,\mathbf{y}_2,y_3) : \begin{array}{c} r_{k+1} \le y_1 - x_1 < r_k \\ |\mathbf{y}_2 - \mathbf{x}_2| < \sqrt{r_k^2 - r_{k+1}^2} \\ |y_3 - x_3| < h^*(x_1, r_k) \end{array} \right\}. \tag{8.13}$$

Note the estimate

$$\begin{aligned} |\mathbf{y}_2 - \mathbf{x}_2| &< \sqrt{r_k^2 - r_{k+1}^2} \approx \sqrt{r_k - r_{k+1}}\sqrt{r_k} \\ \text{for } y &\in \widetilde{E}(x,r_k). \end{aligned} \tag{8.14}$$

We claim the following lemma.

LEMMA 60. *With notation as above we have*

$$\left|\widetilde{E}(x,r_k)\right| \approx |E(x,r_k)| \approx |E(x,r_k) \cap B(x,r_k)| \approx |B(x,r_k)|.$$

PROOF. Recall that by Lemma 57 we have

$$|B(x,r_k)| \approx \begin{cases} r_k^n f(x_1) & \text{if } \quad r_k \le \frac{2}{|F'(x_1)|} \\ \frac{f(x_1+r_k)}{|F'(x_1+r_k)|^n} \left(r_k |F'(x_1+r_k)|\right)^{\frac{n}{2}-1} & \text{if } \quad r_k \ge \frac{2}{|F'(x_1)|} \end{cases}. \tag{8.15}$$

To show

$$|E(x,r_k)| \approx |B(x,r_k)| \tag{8.16}$$

we first note that $|E(x,r_k)| \approx \left|\widetilde{E}(x,r_k)\right|$. Integrating, we easily obtain

$$|\widetilde{E}(x,r_k)| \approx h^*(x_1,r_k)\, r_k^{\frac{n-2}{2}} (r_k - r_{k+1})^{\frac{n}{2}}.$$

Now, in the first case $r_k \geq \frac{1}{|F'(x_1)|}$ we have by Proposition 47 $r_k - r_{k+1} = r_k - r_k^* \approx \frac{1}{|F'(x_1+r_k)|}$ and

$$h^*(x_1,r_k) = h^*\left(x_1,r^*_{k-1}\right) = h(x_1,r_{k-1}) \approx \frac{f(x_1+r_{k-1})}{|F'(x_1+r_{k-1})|} \approx \frac{f(x_1+r_k)}{|F'(x_1+r_k)|},$$

where for the last set of inequalities we used $r_k = r^*_{k-1}$ and the estimate (7.12). This gives

$$|\widetilde{E}(x,r_k)| \approx r_k^{\frac{n}{2}-1} \frac{f(x_1+r_k)}{|F'(x_1+r_k)|^{\frac{n}{2}+1}},$$

which is the second estimate in (8.15) provided we also have $r_k \geq \frac{2}{|F'(x_1)|}$. Moreover, when $\frac{1}{|F'(x_1)|} \leq r_k \leq \frac{2}{|F'(x_1)|}$ the two estimates in (8.15) coincide, so we conclude (8.16) for $r_k \geq \frac{1}{|F'(x_1)|}$.

In the second case $r_k < \frac{1}{|F'(x_1)|}$ we have $r_k - r_{k+1} = \frac{r_k}{2}$ and using part (3) of Proposition 47

$$h^*(x_1,r_k) \approx h^*(x_1,r_{k+1}) = h(x_1,r_k) \approx r_k f(x_1),$$

which gives

$$|\widetilde{E}(x,r_k)| \approx r_k^n f(x_1).$$

This concludes the proof of (8.16).

We are thus left to show

$$|E(x,r_k) \cap B(x,r_k)| \approx |B(x,r_k)|. \tag{8.17}$$

In the case $r_k \geq \frac{1}{|F'(x_1)|}$ we have

$$E(x,r_k) \cap B(x,r_k) \supset B_+(x,r_k) \equiv \{(y_1,\mathbf{y}_2,y_3) \in B(x,r_k) : y_1 > x_1 + r_k^*\}$$

and therefore

$$\begin{aligned}
&|E(x,r_k)\cap B(x,r_k)| \geq |B_+(x,r_k)| \\
&= \int_{|\mathbf{y}_2|\leq r_k} \left| B_{2D}\left((x_1,\mathbf{0},0), \sqrt{r_k^2-|\mathbf{y}_2|^2}\right) \cap \{y_1 - x_1 > r_k^*\}\right| d\mathbf{y}_2 \\
&\approx \int_{|\mathbf{y}_2|\leq r_k} \left(\sqrt{r_k^2-|\mathbf{y}_2|^2} - r_k^*\right)_+ \cdot h\left(x_1, \sqrt{r_k^2-|\mathbf{y}_2|^2}\right) d\mathbf{y}_2 \\
&\approx \int_{|\mathbf{y}_2|\leq r_k} \left(\sqrt{r^2-|\mathbf{y}_2|^2} - r_k^*\right)_+ \cdot \frac{f\left(x_1+\sqrt{r_k^2-|\mathbf{y}_2|^2}\right)}{\left|F'\left(x_1+\sqrt{r^2-|\mathbf{y}_2|^2}\right)\right|} d\mathbf{y}_2 \\
&= \int_{|\mathbf{y}_2|^2\leq r_k^2-r_k^{*2}} \left(\sqrt{r_k^2-|\mathbf{y}_2|^2} - r_k^*\right) \cdot \frac{f\left(x_1+\sqrt{r^2-|\mathbf{y}_2|^2}\right)}{\left|F'\left(x_1+\sqrt{r_k^2-|\mathbf{y}_2|^2}\right)\right|} d\mathbf{y}_2 \\
&\approx \frac{f(x_1+r_k)}{|F'(x_1+r_k)|} \int_{|\mathbf{y}_2|^2\leq r_k^2-r_k^{*2}} \left(\sqrt{r_k^2-|\mathbf{y}_2|^2} - r_k^*\right) d\mathbf{y}_2,
\end{aligned}$$

where for the last equality we used (7.12). Passing to the polar coordinates, $\rho = |\mathbf{y}_2|$, we have

$$\begin{aligned}
\int_{|\mathbf{y}_2|^2\leq r_k^2-r_k^{*2}} \left(\sqrt{r_k^2-|\mathbf{y}_2|^2} - r_k^*\right) d\mathbf{y}_2 &\approx \int_0^{\sqrt{r_k^2-r_k^{*2}}} \left(\sqrt{r_k^2-\rho^2} - r_k^*\right) \rho^{n-3} d\rho \\
&\geq \int_0^{\sqrt{r_k^2-\frac{1}{4}(r_k+r_k^*)^2}} \left(\sqrt{r_k^2-\rho^2} - r_k^*\right) \rho^{n-3} d\rho \\
&\geq \frac{1}{2}(r_k - r_k^*) \int_0^{\sqrt{r_k^2-\frac{1}{4}(r_k+r_k^*)^2}} \rho^{n-3} d\rho \\
&\approx (r_k - r_k^*)^{\frac{n}{2}} r^{\frac{n}{2}-1} \\
&\approx (r_k - r_k^*)^{\frac{n}{2}} r_k^{\frac{n}{2}-1}.
\end{aligned}$$

Using part (2) of Proposition 47 we have $r_k - r_k^* \approx 1/|F'(x_1+r_k)|$ and therefore

$$|E(x,r_k)\cap B(x,r_k)| \gtrsim \frac{f(x_1+r_k)}{|F'(x_1+r_k)|^{\frac{n}{2}+1}} r^{\frac{n}{2}-1} \approx |B(x,r_k)|.$$

Finally, in the case $r_k < \frac{1}{|F'(x_1)|}$, proceeding exactly as in the proof of (8.4) in the 2-dimensional case, we can show

$$E(x,r_k)\cap B(x,r_k) \supset \left[x_1+\frac{r_k}{2}, x_1+\frac{3r_k}{4}\right] \times \left\{|\mathbf{y}_2| \leq \frac{r_k}{2}\right\} \times [-cf(x_1)r_k, cf(x_1)r_k]$$

and thus

$$|E(x,r_k)\cap B(x,r_k)| \gtrsim r_k^n f(x_1) \approx |B(x,r_k)|.$$

This concludes the proof of (8.17), and therefore the proof of Lemma 60. □

1.2.1. *The difficulty with the standard sequence of radii.* Recall that we began the proof of Lemma 58 in two dimensions by subtracting consecutive averages of w over the ends $E(x,r_k)$ and $E(x,r_{k+1})$ to obtain

$$\begin{aligned}
w(x) - &\mathbb{E}_{x,r_1} w \\
&= \lim_{k\to\infty} \frac{1}{|E(x,r_k)|}\int_{E(x,r_k)} w(y)\,dy - \mathbb{E}_{x,r_1} w \\
&= \sum_{k=1}^{\infty}\left\{\frac{1}{|E(x,r_{k+1})|}\int_{E(x,r_{k+1})} w(y)\,dy - \frac{1}{|E(x,r_k)|}\int_{E(x,r_k)} w(z)\,dz\right\}.
\end{aligned}$$

If we simply proceed in this way in higher dimensions we will obtain, just as in the two dimensional proof, that

$$\begin{aligned}
&|w(x) - \mathbb{E}_{x,r_1} w| \\
&\lesssim \sum_{k=1}^{\infty}\frac{1}{|B(x,r_k)|^2}\int_{E(x,r_{k+1})\times E(x,r_k)} |w(y)-w(z)|\,dydz \\
&\lesssim \sum_{k=1}^{\infty}\frac{1}{|B(x,r_k)|^2}\int_{E(x,r_{k+1})\times E(x,r_k)} |w(y_1,\mathbf{y}_2,y_3)-w(z_1,\mathbf{y}_2,y_3)|\,dydz \\
&\quad + \sum_{k=1}^{\infty}\frac{1}{|B(x,r_k)|^2}\int_{E(x,r_{k+1})\times E(x,r_k)} |w(z_1,\mathbf{y}_2,y_3)-w(z_1,\mathbf{z}_2,y_3)|\,dydz \\
&\quad + \sum_{k=1}^{\infty}\frac{1}{|B(x,r_k)|^2}\int_{E(x,r_{k+1})\times E(x,r_k)} |w(z_1,\mathbf{z}_2,y_3)-w(z_1,\mathbf{z}_2,z_3)|\,dydz \\
&\equiv I + II + III,
\end{aligned}$$

but where now

$$I \lesssim \sum_{k=1}^{\infty}\frac{1}{|B(x,r_k)|^2}\int_{E(x,r_{k+1})\times E(x,r_k)}\int_{y_1}^{z_1} |w_{x_1}(s,\mathbf{y}_2,y_3)|\,dsdy_1d\mathbf{y}_2dy_3dz_1d\mathbf{z}_2dz_3\ ,$$

and

$$\begin{aligned}
II \lesssim \sum_{k=1}^{\infty}\frac{1}{|B(x,r_k)|^2}\int_{E(x,r_{k+1})\times E(x,r_k)}\int_0^1 &|(\mathbf{z}_2-\mathbf{y}_2)\cdot\nabla_{\mathbf{x}_2} w(z_1,t\mathbf{z}_2+(1-t)\mathbf{y}_2,y_3)| \\
&\times dtdy_1d\mathbf{y}_2dy_3dz_1d\mathbf{z}_2dz_3\ ,
\end{aligned}$$

and

$$III \lesssim \sum_{k=1}^{\infty}\frac{1}{|B(x,r_k)|^2}\int_{E(x,r_{k+1})\times E(x,r_k)}\int_{y_3}^{z_3} |w_{x_3}(z_1,\mathbf{z}_2,u)|\,dudy_1d\mathbf{y}_2dy_3dz_1d\mathbf{z}_2dz_3\ .$$

Thus with $\triangle r_k \equiv r_k - r_{k+1}$, we have

$$I \lesssim \sum_{k=1}^{\infty} \frac{\triangle r_k}{|B(x,r_k)|} \int_{H(x,r_k)} |w_{x_1}(s,\mathbf{y}_2,y_3)|\, ds d\mathbf{y}_2 dy_3$$
$$\lesssim \sum_{k=1}^{\infty} \frac{\triangle r_k}{|B(x,r_k)|} \int_{H(x,r_k)} |\nabla_A w|,$$

and an easy computation also shows that

$$III \lesssim \sum_{k=1}^{\infty} \frac{\triangle r_k}{|B(x,r_k)|} \int_{H(x,r_k)} |\nabla_A w|,$$

which for terms I and III delivers the good estimate (8.7) in the 2-dimensional proof above. But upon using the inequality $|\mathbf{y}_2 - \mathbf{x}_2| \lesssim \sqrt{r_k - r_{k+1}}\sqrt{r_k}$ from (8.14), the corresponding estimate for II is

$$II \lesssim \sum_{k=1}^{\infty} \frac{\sqrt{r_k \triangle r_k}}{|B(x,r_k)|} \int_{H(x,r_k)} |\nabla_A w|,$$

which is much too large as $\sqrt{r_k \triangle r_k} \gg \triangle r_k$ when $\triangle r_k \ll r_k$.

This suggests that we hold the variable $\mathbf{y}_2$ fixed, and take the difference of lower dimensional averages, and then average over $\mathbf{y}_2$. But this will require additional information on the regularity of the sequence of radii $\{r_k\}_{k=1}^{\infty}$, something we cannot easily derive from the current definition of r_k. So we now turn to redefining the sequence of radii to be used in our subrepresentation inequalities.

1.2.2. *Geometric estimates.* We will estimate the differences of the quantities

$$q_k \equiv \sqrt{r_k^2 - r_{k+1}^2} \tag{8.18}$$

appearing as the widths of the modified ends $\widetilde{E}(x_1,r_k)$ defined in (8.13). First recall that there are positive constants c, C such that

$$c \le (r_k - r_{k+1})\,|F'(x_1+r_k)| \le C, \qquad \text{for } r_k \ge \frac{2}{|F'(x_1)|}.$$

In view of this, let us redefine, for each $\gamma > 0$, the sequence $\{r_k\}_{k=0}^{\infty}$ recursively by demanding that the first inequality above be an equality.

DEFINITION 61. For $\gamma > 0$ set

$$r_{k+1}^{\gamma} \equiv \begin{cases} r_k^{\gamma} - \frac{\gamma}{|F'(x_1+r_k)|} & \text{if} \quad r_k^{\gamma} \ge \frac{\gamma}{|F'(x_1)|} \\ \frac{1}{2} r_k^{\gamma} & \text{if} \quad r_k^{\gamma} < \frac{\gamma}{|F'(x_1)|} \end{cases}, \qquad k \ge 0. \tag{8.19}$$

We will typically suppress the superscript γ and continue to write r_k in place of r_k^{γ} when γ is understood. With this revised definition of the sequence of radii, and the corresponding balls and ends, we retain the two-dimensional volume estimates and the subrepresentation inequality in Lemma 58, with perhaps larger constants of comparability. These details are easily verified and left for the reader.

Now, continuing to suppress γ, let

$$\triangle r_k \equiv r_k - r_{k+1} \text{ and } \triangle^2 r_k \equiv \triangle r_k - \triangle r_{k+1} \text{ and } \triangle q_k \equiv q_k - q_{k+1}$$

denote the first and second order differences of the sequences $\{r_k\}_{k=0}^{\infty}$ and $\{q_k\}_{k=0}^{\infty}$. The point of the new definitions of the sequence $\{r_k^{\gamma}\}_{k=0}^{\infty}$ is to obtain a good estimate on its second order differences $\triangle^2 r_k^\gamma$, and hence also on the first order differences of $\{q_k^\gamma\}_{k=0}^\infty$.

LEMMA 62. *With $\gamma > 0$ and the sequence $\{r_k^\gamma\}_{k=0}^\infty$ defined as in (8.19), and $\{q_k^\gamma\}_{k=0}^\infty$ defined as in (8.18), we have the following estimates:*

$$\begin{aligned} \triangle r_k^\gamma &= \begin{cases} \frac{\gamma}{|F'(x_1+r_k)|} & \text{if} \quad r_k \geq \frac{\gamma}{|F'(x_1)|} \\ \frac{1}{2} r_k & \text{if} \quad r_k < \frac{\gamma}{|F'(x_1)|} \end{cases}, \\ \triangle^2 r_k^\gamma &\approx \frac{[\triangle r_k]^2}{r_k}, \\ q_k^\gamma &\approx \sqrt{r_k \triangle r_k}, \\ \triangle q_k^\gamma &\lesssim \triangle r_k, \end{aligned}$$

where the implied constants of comparability depend on $\gamma > 0$.

PROOF. We suppress the superscript γ and prove only the case where $r_k \geq \frac{\gamma}{|F'(x_1)|}$, since in the opposite case where $r_{k+1} = \frac{1}{2} r_k$ and $\triangle r_k = \frac{1}{2} r_k$, the estimates then follow immediately. We begin with

$$\begin{aligned} \triangle^2 r_k &= \triangle r_k - \triangle r_{k+1} = \frac{\gamma}{|F'(x_1 + r_k)|} - \frac{\gamma}{|F'(x_1 + r_{k+1})|} \\ &= \gamma \frac{F'(x_1 + r_k) - F'(x_1 + r_{k+1})}{|F'(x_1 + r_k)| \, |F'(x_1 + r_{k+1})|} \\ &= \gamma \frac{F''(x_1 + (1-\theta) r_k + \theta r_{k+1}) \, (r_k - r_{k+1})}{|F'(x_1 + r_k)| \, |F'(x_1 + r_{k+1})|} \\ &\approx \frac{\frac{|F'(x_1 + (1-\theta) r_k + \theta r_{k+1})|}{x_1 + (1-\theta) r_k + \theta r_{k+1}} (r_k - r_{k+1})}{|F'(x_1 + r_k)| \, |F'(x_1 + r_{k+1})|} \lesssim \frac{[\triangle r_k]^2}{r_{k+1}}, \end{aligned}$$

where the last two line follows from our assumptions on F and (8.19). Next we have

$$q_k = \sqrt{r_k - r_{k+1}} \sqrt{r_k + r_{k+1}} \approx \sqrt{r_k \triangle r_k},$$

and finally we have

$$\begin{aligned} \triangle q_k &= \sqrt{\triangle r_k} \sqrt{r_k + r_{k+1}} - \sqrt{\triangle r_{k+1}} \sqrt{r_{k+1} + r_{k+2}} \\ &= \left(\sqrt{\triangle r_k} - \sqrt{\triangle r_{k+1}} \right) \sqrt{r_k + r_{k+1}} \\ &\quad + \sqrt{\triangle r_{k+1}} \left(\sqrt{r_k + r_{k+1}} - \sqrt{r_{k+1} + r_{k+2}} \right) \\ &\equiv I + II. \end{aligned}$$

Now

$$\begin{aligned} I &\approx \left(\sqrt{\triangle r_k} - \sqrt{\triangle r_{k+1}} \right) \sqrt{r_k} \approx \left(\frac{\triangle r_k - \triangle r_{k+1}}{\sqrt{\triangle r_k} + \sqrt{\triangle r_{k+1}}} \right) \sqrt{r_k} \\ &\approx \left(\triangle^2 r_k \right) \sqrt{\frac{r_k}{\triangle r_k}} \lesssim \frac{[\triangle r_k]^2}{r_{k+1}} \sqrt{\frac{r_k}{\triangle r_k}} = \frac{r_k}{r_{k+1}} \sqrt{\frac{\triangle r_k}{r_k}} \triangle r_k \lesssim \triangle r_k \end{aligned}$$

and

$$\begin{aligned} II &= \sqrt{\triangle r_{k+1}}\frac{(r_k+r_{k+1})-(r_{k+1}+r_{k+2})}{\sqrt{r_k+r_{k+1}}+\sqrt{r_{k+1}+r_{k+2}}} \\ &\approx \sqrt{\frac{\triangle r_{k+1}}{r_k}}\,\triangle\, r_k \lesssim \triangle r_k\,. \end{aligned}$$

□

1.2.3. *Statements of subrepresentation inequalities.* Set

$$\mathbb{E}_{x,r_1}w\equiv\frac{1}{|E(x,r_1)|}\int_{E(x,r_1)}w.$$

LEMMA 63 (nD subrepresentation). *With $\Gamma(x,r)$ as above, and $r_0=r$ and r_1 given by (8.1), we have the subrepresentation formula*

$$|w(x)-\mathbb{E}_{x,r_1}w|\le C\int_{\Gamma(x,r)}|\nabla_A w(y)|\frac{\widehat{d}(x,y)}{|B(x,d(x,y))|}dy,$$

where ∇_A is as in (1.4) and

$$\widehat{d}(x,y)\equiv\min\left\{d(x,y),\frac{1}{|F'(x_1+d(x,y))|}\right\}.$$

We next claim that the subrepresentation inequality continues to hold if we use the *modified* 'end' $\widetilde{E}(x,r_k)$ in (8.13) to define a *modified* cusp-like region

$$\widetilde{\Gamma}(x,r)=\bigcup_{k=1}^{\infty}\widetilde{E}(x,r_k)\,. \tag{8.20}$$

Indeed, Lemma 63 extends to higher dimensions with $\widetilde{\Gamma}(x,r)$ in place of $\Gamma(x,r)$ in the subrepresentation formula, and with the average

$$\widetilde{\mathbb{E}}_{x,r_1}w\equiv\frac{1}{\left|\widetilde{E}(x,r_1)\right|}\int_{\widetilde{E}(x,r_1)}w$$

in place of $\mathbb{E}_{x,r_1}w$.

LEMMA 64 (nD subrepresentation). *With notation as above, and $r_0=r$, r_1 given by (8.1), and $\widetilde{\Gamma}$ by (8.20), we have the subrepresentation formula*

$$\left|w(x)-\widetilde{\mathbb{E}}_{x,r_1}w\right|\le C\int_{\widetilde{\Gamma}(x,r)}|\nabla_A w(y)|\frac{\widehat{d}(x,y)}{|B(x,d(x,y))|}dy,$$

where ∇_A is as in (1.4) and

$$\widehat{d}(x,y)\equiv\min\left\{d(x,y),\frac{1}{|F'(x_1+d(x,y))|}\right\}.$$

Combining Lemma 64 with Lemma 57, we obtain that in dimension $n \geq 3$ we have the estimate

(8.21)

$$K_{B(0,r_0)}(x,y) = \frac{\widehat{d}(x,y)}{|B(x,d(x,y))|}\mathbf{1}_{\widetilde{\Gamma}(x,r_0)}(y)$$
$$\approx \begin{cases} \dfrac{1}{r^{n-1}f(x_1)}\mathbf{1}_{\widetilde{\Gamma}(x,r_0)}(y), & 0 < r = y_1 - x_1 < \dfrac{2}{|F'(x_1)|} \\ \dfrac{|F'(x_1+r)|^{n-1}}{f(x_1+r)\lambda(x_1,r)^{n-2}}\mathbf{1}_{\widetilde{\Gamma}(x,r_0)}(y), & R \geq r = y_1 - x_1 \geq \dfrac{2}{|F'(x_1)|} \end{cases},$$

where we have defined

$$\lambda(x_1,r) \equiv \sqrt{r|F'(x_1+r)|}. \tag{8.22}$$

The proofs of Lemmas 63 and 64 are both similar to that of the two dimensional analogue, Lemma 58 above. The main difference lies in the fact, already noted above, that we can no longer simply subtract the averages of w over the ends $\widetilde{E}(x,r_k)$ and $\widetilde{E}(x,r_{k+1})$ since the diameter of these ends in the $\mathbf{x}_2$ direction is comparable to $\sqrt{r_k(r_k - r_{k+1})}$, a quantity much larger than $r_k - r_{k+1}$ when $r_k - r_{k+1} \ll r_k$.

There is one more estimate we give. Define the half metric ball

$$HB(0,r) = B(0,r) \cap \{(x,y) \in \mathbb{R}^2 : x > 0\}.$$

We show that for $x \in HB(0,r)$ and $0 < x_1 \leq r^* = r^*(0,r)$, the rectangle $\widetilde{E}(x,r-r^*)$ has volume comparable to that of $\widetilde{E}(0,r)$, and hence the averages of w over the modified ends $\widetilde{E}(x,r-r^*)$ and $\widetilde{E}(0,r)$ have controlled difference. We will not use this estimate in this paper, but it is natural and may prove useful elsewhere. More precisely we have the following lemma.

LEMMA 65. *Suppose that $0 < r < R$, $x \in HB(0,r)$, and $0 < x_1 \leq r^*$. Then we have*

(1) $\left|\widetilde{E}(x,r-r^*)\right| \approx \left|\widetilde{E}(x,r-r^*) \cap HB(0,r)\right| \approx \left|\widetilde{E}(0,r)\right|$,

(2) $\left|\frac{1}{|\widetilde{E}(x,r-r^*)|}\int_{\widetilde{E}(x,r-r^*)\cap HB(0,r)} w - \frac{1}{|\widetilde{E}(0,r)|}\int_{\widetilde{E}(0,r)} w\right| \lesssim r\int_{HB(0,r)}|\nabla_A w|dx$.

PROOF. This is a straightforward exercise. □

As a consequence of this lemma, we obtain the average control

$$\left|\frac{1}{\left|\widetilde{E}(x,r-r^*)\right|}\int_{\widetilde{E}(x,r-r^*)\cap B_+(0,r)} w - \frac{1}{\left|\widetilde{E}(0,r)\right|}\int_{\widetilde{E}(0,r)} w\right| \lesssim r\int_{B(0,r)}|\nabla_A w|dx. \tag{8.23}$$

For the case when $r^* < x_1 \leq r$, we simply use instead of the end $\widetilde{E}(0,r)$, the substitute

$$\overleftrightarrow{E}(0,r) \equiv \left\{ y = (y_1,\mathbf{y}_2,y_3) : \begin{array}{c} r - 3(r-r^*) \leq y_1 - x_1 < r - 2(r-r^*) \\ |\mathbf{y}_2 - \mathbf{x}_2| < \sqrt{r_k^2 - (y_1-x_1)^2} \\ |y_3 - x_3| < h^*(x_1, y_1 - x_1) \end{array} \right\},$$

which looks like $\widetilde{E}(0,r)$ translated a distance $2(r-r^*)$ to the left. This gives the average control (8.23) in the case $r^* < x_1 \leq r$ as well. With these considerations we have obtained the following corollary.

COROLLARY 66. *With notation as above, and* $r_0 = r$, r_1 *given by* (8.1), *suppose that* $\widetilde{\mathbb{E}}_{0,r} w = 0$. *Then we have the subrepresentation formula*

$$|w(x)| \leq C \int_{\widetilde{\Gamma}(x,r)} |\nabla_A w(y)| \frac{\widehat{d}(x,y)}{|B(x,d(x,y))|} dy,$$

where ∇_A *is as in* (1.4) *and*

$$\widehat{d}(x,y) \equiv \min\left\{ d(x,y), \frac{1}{|F'(x_1 + d(x,y))|} \right\}.$$

1.2.4. *Proofs of the subrepresentation inequalities.* We begin with a preliminary estimate on difference of averages that will set the stage for the proofs of the subrepresentation inequalities. Recall that the modified end $\widetilde{E}(x,r_k)$ defined in (8.13) is a product set consisting of those $y = (y_1, \mathbf{y}_2, y_3) \in \mathbb{R} \times \mathbb{R}^{n-2} \times \mathbb{R} = \mathbb{R}^n$ belonging to

$$\widetilde{E}(x,r_k) = [x_1 + r_{k+1}, x_1 + r_k) \times Q(\mathbf{x}_2, q_k) \times (x_3 - h^*(x_1,r_k), x_3 + h^*(x_1,r_k)),$$

where $Q(\mathbf{z},q)$ denotes the $(n-2)$-dimensional Euclidean ball centered at $\mathbf{z} \in \mathbb{R}^{n-2}$ with radius q. With Lemma 62 in hand, we now dilate the modified end $\widetilde{E}(x,r_k)$ in the $\mathbf{x}_2$ variable so that it has the same 'thickness' as $\widetilde{E}(x,r_{k+1})$, namely q_{k+1}. More precisely, we define

$$\widehat{E}(x,r_k) \equiv [x_1 + r_{k+1}, x_1 + r_k) \times Q(\mathbf{x}_2, q_{k+1}) \times (x_3 - h^*(x_1,r_k), x_3 + h^*(x_1,r_k)). \tag{8.24}$$

Note that the only difference between $\widehat{E}(x,r_k)$ and $\widetilde{E}(x,r_k)$ is the change of width from q_k to q_{k+1}. Then we observe that the dilation

$$\mathbf{y}_2 \to \mathbf{y}_2' = \mathbf{x}_2 + \frac{q_{k+1}}{q_k}(\mathbf{y}_2 - \mathbf{x}_2),$$

with y_1 and y_3 kept fixed, maps $\widetilde{E}(x,r_k)$ one-to-one onto $\widehat{E}(x,r_k)$, and satisfies

$$\begin{aligned}
&\frac{1}{\left|\widetilde{E}(x,r_k)\right|}\int_{\widetilde{E}(x,r_k)} w(y)\,dy - \frac{1}{\left|\widehat{E}(x,r_k)\right|}\int_{\widehat{E}(x,r_k)} w(y)\,dy \\
&= \frac{1}{\triangle r_k}\int_{r_{k+1}}^{r_k}\frac{1}{2h^*(x_1,r_k)} \\
&\qquad \times \int_{-h^*(x_1,r_k)}^{h^*(x_1,r_k)}\left[\frac{1}{|Q(\mathbf{x}_2,q_k)|}\int_{\mathbf{y}_2\in Q(\mathbf{x}_2,q_k)} w(y_1,\mathbf{y}_2,y_3)\,d\mathbf{y}_2\right]dy_3dy_1 \\
&- \frac{1}{\triangle r_k}\int_{r_{k+1}}^{r_k}\frac{1}{2h^*(x_1,r_k)} \\
&\qquad \times \int_{-h^*(x_1,r_k)}^{h^*(x_1,r_k)}\left[\frac{1}{|Q(\mathbf{x}_2,q_{k+1})|}\int_{\mathbf{y}_2'\in Q(\mathbf{x}_2,q_{k+1})} w(y_1,\mathbf{y}_2',y_3)\,d\mathbf{y}_2'\right]dy_3dy_1,
\end{aligned}$$

where the difference of averages in square brackets over $Q(\mathbf{x}_2,q_k)$ and $Q(\mathbf{x}_2,q_{k+1})$ satisfies

$$\begin{aligned}
&\frac{1}{|Q(\mathbf{x}_2,q_k)|}\int_{\mathbf{y}_2\in B(\mathbf{x}_2,q_k)} w(y_1,\mathbf{y}_2,y_3)\,d\mathbf{y}_2 \\
&\qquad - \frac{1}{|Q(\mathbf{x}_2,q_{k+1})|}\int_{\mathbf{y}_2'\in Q(\mathbf{x}_2,q_{k+1})} w(y_1,\mathbf{y}_2',y_3)\,d\mathbf{y}_2' \\
&= \frac{1}{|Q(\mathbf{x}_2,q_k)|}\int_{\mathbf{y}_2'\in Q(\mathbf{x}_2,q_{k+1})} w\left(y_1,\mathbf{x}_2+\frac{q_k}{q_{k+1}}(\mathbf{y}_2'-\mathbf{x}_2),y_3\right)\frac{\partial\mathbf{y}_2}{\partial\mathbf{y}_2'}d\mathbf{y}_2' \\
&\qquad - \frac{1}{|Q(\mathbf{x}_2,q_{k+1})|}\int_{\mathbf{y}_2'\in Q(\mathbf{x}_2,q_{k+1})} w(y_1,\mathbf{y}_2',y_3)\,d\mathbf{y}_2' \\
&= \frac{1}{|Q(\mathbf{x}_2,q_{k+1})|}\int_{Q(\mathbf{x}_2,q_{k+1})}\left[w\left(y_1,\mathbf{x}_2+\frac{q_{k+1}}{q_k}(\mathbf{y}_2'-\mathbf{x}_2),y_3\right)-w(y_1,\mathbf{y}_2',y_3)\right]d\mathbf{y}_2'.
\end{aligned}$$

Thus we have the following estimate for the difference of averages

$$\begin{aligned}
\widetilde{\mathbb{E}}_{x,r_k} &\equiv \frac{1}{\left|\widetilde{E}(x,r_k)\right|}\int_{\widetilde{E}(x,r_k)} w(y)\,dy, \\
\widehat{\mathbb{E}}_{x,r_k} &\equiv \frac{1}{\left|\widehat{E}(x,r_k)\right|}\int_{\widehat{E}(x,r_k)} w(y)\,dy
\end{aligned}$$

over the modified ends $\widehat{E}(x,r_k)$ and $E(x,r_k)$:

$$\begin{aligned}
\left|\widetilde{\mathbb{E}}_{x,r_k}-\widehat{\mathbb{E}}_{x,r_k}\right| &\leq \frac{1}{\triangle r_k}\int_{r_{k+1}}^{r_k}\frac{1}{2h^*(x_1,r_k)}\int_{-h^*(x_1,r_k)}^{h^*(x_1,r_k)}\frac{1}{|Q(\mathbf{x}_2,q_{k+1})|} \\
&\times\int_{Q(\mathbf{x}_2,q_{k+1})}\left|w\left(y_1,\mathbf{x}_2+\frac{q_{k+1}}{q_k}(\mathbf{y}_2'-\mathbf{x}_2),y_3\right)-w(y_1,\mathbf{y}_2',y_3)\right|d\mathbf{y}_2'dy_1dy_3 \\
&\leq \frac{1}{\triangle r_k}\int_{r_{k+1}}^{r_k}\frac{1}{2h^*(x_1,r_k)}\int_{-h^*(x_1,r_k)}^{h^*(x_1,r_k)}\frac{1}{|Q(\mathbf{x}_2,q_{k+1})|} \\
&\times\int_{Q(\mathbf{x}_2,q_{k+1})}|\nabla_{\mathbf{x}_2}w(y_1,\mathbf{z}_2,y_3)|\,d\mathbf{z}_2dy_1dy_3\left(1-\frac{q_{k+1}}{q_k}\right)|\mathbf{y}_2'-\mathbf{x}_2| \\
&\leq \triangle q_k\frac{1}{\left|\widetilde{E}(x,r_k)\right|}\int_{\widetilde{E}(x,r_k)}|\nabla_A w(y)|\,dy \lesssim \frac{\triangle r_k}{\left|\widetilde{E}(x,r_k)\right|}\int_{\widetilde{E}(x,r_k)}|\nabla_A w(y)|\,dy.
\end{aligned}$$

This estimate,

$$\left|\widetilde{\mathbb{E}}_{x,r_k}-\widehat{\mathbb{E}}_{x,r_k}\right| \lesssim \frac{\triangle r_k}{\left|\widetilde{E}(x,r_k)\right|}\int_{\widetilde{E}(x,r_k)}|\nabla_A w(y)|\,dy, \tag{8.25}$$

for the difference of averages has the same form as the corresponding estimates for the summands in terms I and III in the previous subsubsection, and thus we can replace the average of w over $\widetilde{E}(x,r_k)$ by its average over $\widehat{E}(x,r_k)$ whenever we wish.

We can now prove the subrepresentation formula in Lemma 64. The proof of Lemma 63 is similar and left for the reader.

PROOF OF LEMMA 64. We have

$$\begin{aligned}
w(x)-\widetilde{\mathbb{E}}_{x,r_1}w &= \lim_{k\to\infty}\frac{1}{\left|\widetilde{E}(x,r_k)\right|}\int_{\widetilde{E}(x,r_k)}w(y)\,dy-\widetilde{\mathbb{E}}_{x,r_1}w \\
&=\sum_{k=1}^{\infty}\left\{\frac{1}{\left|\widetilde{E}(x,r_{k+1})\right|}\int_{\widetilde{E}(x,r_{k+1})}w(y)\,dy-\frac{1}{\left|\widetilde{E}(x,r_{k+1})\right|}\int_{\widetilde{E}(x,r_{k+1})}w(z)\,dz\right\},
\end{aligned}$$

and so

$$\begin{aligned}
&w(x)-\widetilde{\mathbb{E}}_{x,r_1}w \\
&\qquad=\sum_{k=1}^{\infty}\frac{1}{\left|\widetilde{E}(x,r_k)\right|}\int_{\widetilde{E}(x,r_k)}w(y)\,dy-\frac{1}{\left|\widehat{E}(x,r_k)\right|}\int_{\widehat{E}(x,r_k)}w(z)\,dz \\
&\qquad+\sum_{k=1}^{\infty}\frac{1}{\left|\widehat{E}(x,r_k)\right|}\int_{\widehat{E}(x,r_k)}w(y)\,dy-\frac{1}{\left|\widetilde{E}(x,r_{k+1})\right|}\int_{E(x,r_{k+1})}w(z)\,dz \\
&\qquad\equiv I+\sum_{k=1}^{\infty}II_k,
\end{aligned}$$

where

$$|I|\lesssim\sum_{k=1}^{\infty}\frac{\triangle r_k}{\left|\widetilde{E}(x,r_k)\right|}\int_{\widetilde{E}(x,r_k)}|\nabla_A w(y)|\,dy,$$

and

$$
\begin{aligned}
II_k =& \frac{1}{\triangle r_k}\int_{r_{k+1}}^{r_k}\frac{1}{2h^*(x_1,r_k)}\int_{-h^*(x_1,r_k)}^{h^*(x_1,r_k)}\frac{1}{|Q(\mathbf{x}_2,q_{k+1})|}\int_{Q(\mathbf{x}_2,q_{k+1})}w \\
&-\frac{1}{\triangle r_{k+1}}\int_{r_{k+2}}^{r_{k+1}}\frac{1}{2h^*(x_1,r_{k+1})}\int_{-h^*(x_1,r_{k+1})}^{h^*(x_1,r_{k+1})}\frac{1}{|Q(\mathbf{x}_2,q_{k+1})|}\int_{Q(\mathbf{x}_2,q_{k+1})}w \\
=& \frac{1}{|Q(\mathbf{x}_2,q_{k+1})|}\int_{Q(\mathbf{x}_2,q_{k+1})}\left\{\frac{1}{\triangle r_k}\int_{r_{k+1}}^{r_k}\frac{1}{2h^*(x_1,r_k)}\int_{-h^*(x_1,r_k)}^{h^*(x_1,r_k)}w\right. \\
&\left.-\frac{1}{\triangle r_{k+1}}\int_{r_{k+2}}^{r_{k+1}}\frac{1}{2h^*(x_1,r_{k+1})}\int_{-h^*(x_1,r_{k+1})}^{h^*(x_1,r_{k+1})}w\right\}.
\end{aligned}
$$

Now the difference of the 2-dimensional integrals in braces has the variable $\mathbf{y}_2$ fixed, and using the two dimensional proof above, the modulus of this difference is easily seen to be controlled by

$$
[\triangle r_k]\frac{1}{\triangle r_k}\int_{r_{k+2}}^{r_k}\frac{1}{2h^*(x_1,r_k)}\int_{-h^*(x_1,r_k)}^{h^*(x_1,r_k)}|\nabla_A w|.
$$

Then averaging over $\mathbf{y}_2\in B(\mathbf{x}_2,q_{k+1})$, we obtain the bound

$$
\begin{aligned}
|II_k| &\lesssim [\triangle r_k]\frac{1}{|B(\mathbf{x}_2,q_{k+1})|}\int_{B(\mathbf{x}_2,q_{k+1})}\frac{1}{\triangle r_k}\int_{r_{k+2}}^{r_k}\frac{1}{2h^*(x_1,r_k)}\int_{-h^*(x_1,r_k)}^{h^*(x_1,r_k)}|\nabla_A w| \\
&\lesssim [\triangle r_k]\frac{1}{\left|\widetilde{H}(x,r_k)\right|}\int_{\widetilde{H}(x,r_k)}|\nabla_A w|,
\end{aligned}
$$

where $\widetilde{H}_k(x,r_k)\equiv\widetilde{E}(x,r_{k+1})\bigcup\widetilde{E}(x,r_k)$. Altogether then we have

$$
\begin{aligned}
&\left|w(x)-\widetilde{\mathbb{E}}_{x,r_1}w\right| \\
&\qquad\lesssim\sum_{k=1}^{\infty}\left\{\frac{\triangle r_k}{\left|\widetilde{E}(x,r_k)\right|}\int_{\widetilde{E}(x,r_k)}|\nabla_A w|+\frac{\triangle r_k}{\left|\widetilde{H}(x_1,r_k)\right|}\int_{\widetilde{H}(x_1,r_k)}|\nabla_A w|\right\}.
\end{aligned}
$$

Moreover, by estimates above we have that $|\widetilde{H}(x,r_k)|\approx|B(x,r_k)|$, and

$$
\begin{aligned}
\left|w(x)-\widetilde{\mathbb{E}}_{x,r_1}w\right| &\lesssim \sum_{k=1}^{\infty}\frac{r_k-r_{k+1}}{|B(x,r_k)|}\left(\int_{\widetilde{H}(x_1,r_k)}|\nabla_A w(y)|\,dy\right) \\
&\lesssim \int_{\widetilde{\Gamma}(x,r)}|\nabla_A w(y)|\left(\sum_{k=1}^{\infty}\frac{r_k-r_{k+1}}{|B(x,r_k)|}\mathbf{1}_{E(x,r_k)}(y)\right)dy.
\end{aligned}
$$

At this point, we have obtained the higher dimensional analogue of the two-dimensional inequality (8.7), just as in Case 1 and Case 2 of the 2-dimensional proof of Lemma 58, and the proof now proceeds exactly as in the 2-dimensional case there. □

2. $(1,1)$-Sobolev and $(1,1)$-Poincaré inequalities

We will give statements and proofs initially in dimension $n=2$, and later extend the results and arguments to higher dimensions. First, we will obtain the following natural $(1,1)$-Sobolev inequality, which is not actually used in this paper, but included for comparison with the standard kernel in Remark 68 below.

LEMMA 67. *For $w \in Lip_0(B(x_0,r))$ and ∇_A a degenerate gradient as above, we have*

$$\int_{B(x_0,r)} |w(x)|\,dx \le Cr \int_{B(x_0,r)} |\nabla_A w(y)|\,dy.$$

PROOF. If $x \in B(x_0,r)$, then w satisfies the hypothesis of Lemma 58 in $B\left(x, 2(C+1)^2 r\right)$ for the constant C as in (8.2). Indeed, let r_k be defined by (8.1) for $x=y$ and $r_0 = 2(C+1)^2 r$, then

$$r_2 \ge \frac{1}{(C+1)^2} r_0 = 2r.$$

Hence, since

$$E(x,r_1) \equiv \{y : x_1 + r_2 \le y_1 < x_1 + r_1,\ |y_2 - x_2| < h^*(x_1, z_1 - x_1)\},$$

we have that $E(x,r_1) \bigcap B(x_0,r) = \emptyset$ so $\int_{E(x,r_1)} w = 0$ so we may apply Lemma 58 in $B\left(x_0, 2\left((C+1)^2+1\right) r\right)$ for all $x \in B(x_0,r)$.

Let $R = 2\left((C+1)^2+1\right) r$. Using the subrepresentation inequality and (8.34) we have

$$\begin{aligned}
\int |w(x)|\ dx &\le \int\int_{\Gamma(x,R)} \frac{\widehat{d}(x,y)}{|B(x,d(x,y))|} |\nabla_A w(y)|\ dydx \\
&= \int\int K_R(x,y)\ |\nabla_A w(y)|\ dydx \\
&= \int\left\{\int K_R(x,y)\,dx\right\}\ |\nabla_A w(y)|\ dy \\
&\approx \int R|\nabla_A w(y)|\ dy \approx r\int |\nabla_A w(y)|\ dy.
\end{aligned}$$

□

REMARK 68. The larger kernel $\widetilde{K}_r(x,y) \equiv \mathbf{1}_{\Gamma(x,r)}(y) \frac{d(x,y)}{|B(x,d(x,y))|}$, with $\widehat{d}$ replaced by d, does **not** in general yield the $(1,1)$ Sobolev inequality. More precisely, the inequality

$$\int\int \widetilde{K}_r(x,y)\ |\nabla_A w(y)|\ dydx \lesssim r\int |\nabla_A w(y)|\ dy, \qquad 0<r\ll 1, \tag{8.26}$$

fails in the case

$$F(x) = \frac{1}{x},\ x>0.$$

To see this take $y_2 = 0$. We now make estimates on the integral

$$\int \widetilde{K}_r(x,y)\ dx \approx \int_{y_1-r}^{y_1}\left\{\int_{y_2-h_{x,y}}^{y_2+h_{x,y}} \frac{1}{h_{x,y}}\frac{d(x,y)}{\widehat{d}(x,y)}dx_2\right\}dx_1, \tag{8.27}$$

where $\widehat{d}(x,y) = \min\left\{d(x,y), \frac{1}{|F'(x_1+d(x,y))|}\right\}$. Consider the region where

$$d(x,y) \ge \frac{1}{|F'(x_1+d(x,y))|} = (x_1+d(x,y))^2. \tag{8.28}$$

In this region we have

$$\frac{d(x,y)}{\widehat{d}(x,y)} = d(x,y)|F'(x_1 + d(x,y))| = \frac{d(x,y)}{(x_1 + d(x,y))^2}.$$

Moreover, since $d(x,y) \leq r$, we have

$$\frac{d(x,y)}{\widehat{d}(x,y)} \geq \frac{r}{(x_1 + r)^2}.$$

On the other hand, we have $d(x,y) \geq y_1 - x_1$ and $d(x,y) \ll 1$, so the condition in (8.28) is guaranteed by $y_1 - x_1 \geq (x_1 + y_1 - x_1)^2$, i.e. $x_1 \leq y_1 - y_1^2$. We then have the following estimate for (8.27):

$$\int \widetilde{K}_r(x,y)dx \gtrsim \int_{y_1 - r}^{y_1 - y_1^2} \frac{r}{(x_1 + r)^2} dx_1 = \frac{r(r - y_1^2)}{y_1(y_1 - y_1^2 + r)}.$$

Therefore, if $y_1 \leq r$, we have

$$\int \widetilde{K}_r(x,y)dx \gtrsim 1,$$

and (8.26) fails for small $r > 0$.

Now we turn to establishing the $(1,1)$-*Poincaré* inequality. For this we will need the following extension of Lemma 79 in [**RSaW**]. Recall the half metric ball

$$HB(0,r) = B(0,r) \cap \{(x,y) \in \mathbb{R}^2 : x > 0\}.$$

PROPOSITION 69. *Let the balls $B(0,r)$ and the degenerate gradient ∇_A be as above. There exists a constant C such that the Poincaré Inequality*

$$\iint_{HB(0,r)} |w - \bar{w}|\, dxdy \leq Cr \iint_{HB(0,r)} |\nabla_A w| dxdy$$

holds for any Lipschitz function w and sufficiently small $r > 0$. Here $\bar{w}$ is the average defined by

$$\bar{w} = \frac{1}{|HB(0,r)|} \iint_{HB(0,r)} w dxdy.$$

2.1. Proof of Poincaré. The left hand side can be estimated by

$$\begin{aligned}
&\iint_{HB(0,r)} |w - \bar{w}|\, dxdy \\
&= \iint_{HB(0,r)} \left| w(x_1,y_1) - \frac{1}{|HB(0,r)|} \iint_{HB(0,r)} w(x_2,y_2) dx_2 dy_2 \right| dx_1 dy_1 \\
&\leq \frac{1}{|HB(0,r)|} \int_{HB(0,r) \times HB(0,r)} |w(x_1,y_1) - w(x_2,y_2)|\, dx_1 dy_1 dx_2 dy_2
\end{aligned}$$

The idea now is to estimate the difference $|w(x_1,y_1) - w(x_2,y_2)|$ by the integral of ∇w along some path. Because the half metric ball is somewhat complicated geometrically, we can simplify the argument by applying the following lemma, sacrificing only the best constant C in the Poincaré inequality.

LEMMA 70. *Let (X,μ) be a measure space. If $\Omega \subset X$ is the disjoint union of 2 measurable subsets $\Omega = \Omega_1 \cup \Omega_2$ so that the measure of the subsets are comparable*

$$\frac{1}{C_1} \leq \frac{\mu(\Omega_1)}{\mu(\Omega_2)} \leq C_1$$

Then there exists a constant $C = C(C_1)$, such that

$$\iint_{\Omega\times\Omega} |w(x)-w(y)|d\mu(x)d\mu(y) \leq C \iint_{\Omega_1\times\Omega_2} |w(x)-w(y)|d\mu(x)d\mu(y). \tag{8.29}$$

for any measurable function w defined on Ω.

PROOF. Define

$$S_{i,j} = \iint_{\Omega_i\times\Omega_j} |w(x)-w(y)|d\mu(x)d\mu(y), \quad i,j=1,2.$$

Since $\Omega = \Omega_1 \cup \Omega_2$, we can rewrite inequality (8.29) as

$$S_{1,1} + 2S_{1,2} + S_{2,2} \leq CS_{1,2}.$$

Now, we compute

$$\begin{aligned}
S_{1,1} &= \frac{1}{\mu(\Omega_2)} \iiint_{\Omega_1\times\Omega_1\times\Omega_2} |[w(x)-w(z)]+[w(z)-w(y)]|\, d\mu(x)d\mu(y)d\mu(z) \\
&\leq \frac{1}{\mu(\Omega_2)} \iiint_{\Omega_1\times\Omega_1\times\Omega_2} |w(x)-w(z)|\, d\mu(x)d\mu(y)d\mu(z) \\
&\quad + \frac{1}{\mu(\Omega_2)} \iiint_{\Omega_1\times\Omega_1\times\Omega_2} |w(y)-w(z)|\, d\mu(x)d\mu(y)d\mu(z) \\
&= \frac{2\mu(\Omega_1)}{\mu(\Omega_2)} \iint_{\Omega_1\times\Omega_2} |w(x)-w(z)|\, d\mu(x)d\mu(z) = \frac{2\mu(\Omega_1)}{\mu(\Omega_2)} S_{1,2}\ ,
\end{aligned}$$

and similarly $S_{2,2} \leq \frac{2\mu(\Omega_2)}{\mu(\Omega_1)} S_{1,2}$. □

We will apply this lemma with

$$\begin{aligned}
\Omega_1 &= B_+ = \{(x,y) \in HB(0,r_0) : r^* \leq x \leq r\}, \\
\Omega_2 &= B_- = \{(x,y) \in HB(0,r_0) : 0 \leq x \leq r^*\},
\end{aligned}$$

where r^*, B_+ and B_- are as in Lemma 53 above. Then from Lemma 53 we have

$$|\Omega_1| \approx |\Omega_2| \approx |B(0,r_0)|.$$

By Lemma 70, the proof of Proposition 69 reduces to the following inequality:

$$\begin{aligned}
I &= \iint_{\Omega_1\times\Omega_2} |w(x_1,y_1)-w(x_2,y_2)|dx_1dy_1dx_2dy_2 \\
&\leq C|HB(0,r_0)|r_0 \iint_{HB(0,r_0)} |\nabla_A w(x,y)|dxdy.
\end{aligned} \tag{8.30}$$

Let $P_1 = (x_1,y_1) \in \Omega_1$ and $P_2 = (x_2,y_2) \in \Omega_2$. We can connect P_1 and P_2 by first traveling vertically and then horizontally. This integral path is completely contained in the half metric ball. This immediately gives an inequality

$$|w(x_1,y_1)-w(x_2,y_2)| \leq \left|\int_{y_1}^{y_2} w_y(x_1,y)dy\right| + \left|\int_{x_1}^{x_2} w_x(x,y_2)dx\right|.$$

As a result, we have

$$
\begin{aligned}
I &= \iint_{\Omega_1\times\Omega_2} |w(x_1,y_1)-w(x_2,y_2)|dx_1dy_1dx_2dy_2 \\
&\le \iint_{\Omega_1\times\Omega_2} \left|\int_{y_1}^{y_2} w_y(x_1,y)dy\right| dx_1dy_1dx_2dy_2 \\
&\quad + \iint_{\Omega_1\times\Omega_2} \left|\int_{x_1}^{x_2} w_x(x,y_2)dx\right| dx_1dy_1dx_2dy_2 \\
&= I_1 + I_2
\end{aligned}
$$

We first estimate the integral

$$
\begin{aligned}
I_1 &= \iint_{\Omega_1\times\Omega_2} |w(x_1,y_1)-w(x_2,y_2)|dx_1dy_1dx_2dy_2 \\
&\le \iint_{\Omega_1\times\Omega_2} \left|\int_{y_1}^{y_2} w_y(x_1,y)dy\right| dx_1dy_1dx_2dy_2
\end{aligned}
$$

where $\Omega_1 = B_+$ and $\Omega_2 = B_-$. We have

$$
\begin{aligned}
I_1 &\le \int_{B_-}\int_{B_+}\int_{y_1}^{y_2} |w_y(x_1,y)|dydx_1dy_1dx_2dy_2 \\
&\le \int_{B_-}\int_{B_+}\int_{y_1}^{y_2} \frac{1}{f(x_1)}|\nabla_A w(x_1,y)|dydx_1dy_1dx_2dy_2 \\
&\le \int_{B_-}\int_{B_+} \frac{h(r)}{f(x_1)}|\nabla_A w(x_1,y)|dydx_1dx_2dy_2,
\end{aligned}
$$

where $h(r) \lesssim rf(r)$ is the "maximal height" given in Proposition 47. Moreover, for $x_1 \in B_+$ we have $|r - x_1| \le 1/|F'(r)|$ and therefore $f(x_1) \approx f(r)$. This gives

$$
\frac{h(r)}{f(x_1)} \le r,
$$

and substituting this into the above we get

$$
I_1 \le Cr|B_-| \int_{B_+} |\nabla_A w(x,y)|dxdy \le Cr|B| \int_B |\nabla_A w(x,y)|dxdy.
$$

To estimate

$$
I_2 = \iint_{\Omega_1\times\Omega_2} \left|\int_{x_1}^{x_2} w_x(x,y_2)dx\right| dx_1dy_1dx_2dy_2,
$$

we note that $|w_x(x,y_2)| \le |\nabla_A w(x,y_2)|$, and therefore

$$
\begin{aligned}
I_2 &\le \iint_{B_+\times B_-} \left[\int_{(x,y_2)\in HB(0,r)} |\nabla_A w(x,y_2)|dx\right] dx_1dy_1dx_2dy_2 \\
&\le Cr|B_+| \int_B |\nabla_A w(x,y_2)|dxdy_2 \le Cr|B| \int_B |\nabla_A w(x,y)|dxdy.
\end{aligned}
$$

This finishes the proof of inequality (8.30), and hence finishes the proof of the Poincaré inequality in Proposition 69.

We now wish to extend this Poincaré inequality to hold for the full ball $B(0,r)$ in Proposition 69. We cannot simply use geodesics that connect the left end $E_{\text{left}}(0,r)$ of the ball to the right end $E_{\text{right}}(0,r)$ of the ball and that also stay entirely within the ball $B(0,r)$. The problem is that the thin 'neck' of the ball near the origin is too thin to support such geodesics without compensating with a huge Jacobian. Instead we will enlarge the ball $B(0,r)$ enough so that the enlarged ball contains the *rectangle* $(-r,r)\times(-h(r),h(r))$. This can be achieved with the ball of doubled radius as we now show.

LEMMA 71. *For $0<r<\frac{R}{2}$, we have the inclusions,*

$$B(0,r)\subset(-r,r)\times(-h(r),h(r))\subset(-r,r)\times(-2h(r),2h(r))\subset B(0,2r).$$

PROOF. The inclusion $B(0,r)\subset(-r,r)\times(-h(r),h(r))$ is immediate. Now consider the geodesic $\gamma(t)$ from $\gamma(0)=(0,0)$ to the point $\gamma(r)=(r^*,h)=(r^*(r),h(r))$ on the boundary of the ball $B(0,r)$ where γ has a vertical tangent. If we continue this geodesic for a further time r, then by symmetry we curl back and return to the y-axis at the point $\gamma(2r)=(0,2h(r))$. It is now clear by a further symmetry that $(-r,r)\times(-h(r),h(r))\subset B(0,2r)$. □

Now we can extend Proposition 69 to the full ball.

PROPOSITION 72. *Let the balls $B(0,r)$ and the degenerate gradient ∇_A be as above. There exists a constant C such that the Poincaré Inequality*

$$\int_{B(0,r)}|w(x)-\bar{w}|\,dx\le Cr\int_{B(0,2r)}|\nabla_A w|dx$$

holds for any Lipschitz function w and sufficiently small $r>0$. Here $\bar{w}$ is the average defined by

$$\bar{w}=\frac{1}{|B(0,r)|}\int_{B(0,r)}wdx.$$

PROOF. Following Proposition 69 we will denote the right half of the metric ball $B(0,r)$, by $HB(0,r)$. Recall that we have $HB(0,r)=B_-\cup B_+$ where

$$\begin{aligned}B_+ &= \{(x_1,x_2)\in HB(0,r_0): r^*\le x_1\le r\},\\ B_- &= \{(x_1,x_2)\in HB(0,r_0): 0\le x_1\le r^*\}.\end{aligned}$$

Similarly, we will denote the left half by $BH(0,r)$ and write $BH(0,r)=B^-\cup B^+$ where

$$\begin{aligned}B^+ &= \{(x_1,x_2)\in BH(0,r_0): -r\le x_1\le -r^*\},\\ B^- &= \{(x_1,x_2)\in BH(0,r_0): -r^*\le x_1\le 0\}.\end{aligned}$$

Now using Lemma 70 with $\Omega_1=BH(0,r)$ and $\Omega_2=HB(0,r)$ we have

$$\begin{aligned}\int_{B(0,r)}|w(x)-\bar{w}|\,dx &\le \frac{1}{|B|}\iint_{B\times B}|w(x)-w(y)|\,dxdy\\ &\le \frac{C}{|B|}\iint_{BH\times HB}|w(x)-w(y)|\,dxdy.\end{aligned}$$

Moreover, since we have $|B_-|=|B^-|\approx|B_+|=|B^+|\approx|B|$, proceeding the same way as in the proof of Lemma 70 we can show

$$\begin{aligned}\iint_{BH\times HB}|w(x)-w(y)|\,dxdy &\le C\Big(\iint_{B^-\times B^+}|w(x)-w(y)|\,dxdy\\ &+\iint_{B_-\times B_+}|w(x)-w(y)|\,dxdy+\iint_{B^+\times B_+}|w(x)-w(y)|\,dxdy\Big)\\ &\equiv I_1+I_2+I_3.\end{aligned}$$

The estimate for I_2 follows from the proof of Proposition 69 , and the estimate for I_1 is shown in exactly the same way. We thus have

$$I_1 + I_2 \leq Cr \int_{B(0,r)} |\nabla_A w| dx. \tag{8.31}$$

To estimate I_3 we connect two points $(x_1, x_2) \in B^+$ and $(y_1, y_2) \in B_+$ by a curve consisting of one vertical and one horizontal segment, which according to Lemma 71 lies entirely in the ball $B(0, 2r)$

$$|w(x_1, x_2) - w(y_1, y_2)| \leq \left| \int_{x_2}^{y_2} w_t(x_1, t) dt \right| + \left| \int_{x_1}^{y_1} w_s(s, y_2) ds \right|.$$

Next, proceeding as in the proof of Proposition 69 we obtain

$$\iint_{B^+ \times B_+} \left| \int_{x_2}^{y_2} w_t(x_1, t) dt \right| dx_1 dx_2 dy_1 dy_2$$
$$\leq |B_+| \int_{B^+} \frac{h(r)}{f(x_1)} |\nabla_A w(x_1, t)| dx_1 dt \leq Cr|B| \int_B |\nabla_A w| dx,$$

where in the last inequality we used the fact that $x_1 \in B^+$ and $y_1 \in B_+$, and therefore

$$\frac{h(r)}{f(x_1)} \approx \frac{h(r)}{f(y_1)} \approx \frac{rf(r)}{f(r)} = Cr.$$

Finally, for the integral along the horizontal segment we have

$$\iint_{B^+ \times B_+} \left| \int_{x_1}^{y_1} w_s(s, y_2) ds \right| dx_1 dx_2 dy_1 dy_2$$
$$\leq Cr \int_{B^+} \int_{-h(r)}^{h(r)} \int_{-r}^{r} |w_s(s, y_2)| \, ds dy_2 dx_1 dy_2 \leq Cr|B| \int_{2B} |\nabla_A w| dx,$$

which gives

$$I_3 \leq Cr \int_{B(0,2r)} |\nabla_A w| dx.$$

Combining with (8.31) finishes the proof. □

2.2. Higher dimensional inequalities. First, we state the following n-dimensional analog of Lemma 71

LEMMA 73. *Define the set $Q(r)$ as follows*

$$Q(r) \equiv \{(y_1, \mathbf{y}_2, y_3) : \ |y_1| < \sqrt{r^2 - |\mathbf{y}_2|^2}, |y_3| < 2h(\sqrt{r^2 - |\mathbf{y}_2|^2}), |\mathbf{y}_2| < r\}.$$

Then we have the following inclusion

$$B(0, r) \subset Q(r) \subset B(0, 2r).$$

The cross section $\mathbf{y}_2 = const$ of the set $Q(r)$ is a rectangle

$$(-\sqrt{r^2 - |\mathbf{y}_2|^2}, \sqrt{r^2 - |\mathbf{y}_2|^2}) \times (-2h(\sqrt{r^2 - |\mathbf{y}_2|^2}), 2h(\sqrt{r^2 - |\mathbf{y}_2|^2}))$$

in the (x_1, x_3) plane.

Now, we first define the "ends" of n-dimensional balls, similar to the sets B_+ and B^+ in 2 dimensions. As before we let $HB(0, r) = \{x \in B(0, r) : \ x_1 > 0\}$, $BH(0, r) = \{x \in B(0, r) : \ x_1 < 0\}$ and

$$
\begin{aligned}
B_+ &= \left\{ (x_1, \mathbf{x}_2, x_3) \in IIB\,(0, r_0) : r^* \le x_1 \le r \right\}, \\
B_- &= \left\{ (x_1, \mathbf{x}_2, x_3) \in HB\,(0, r_0) : 0 \le x_1 \le r^* \right\}.
\end{aligned}
$$

Similarly, we will denote the left half by $BH(0,r)$ and write $BH(0,r) = B^- \cup B^+$ where

$$
\begin{aligned}
B^+ &= \left\{ (x_1, \mathbf{x}_2, x_3) \in BH\,(0, r_0) : -r \le x_1 \le -r^* \right\}, \\
B^- &= \left\{ (x_1, \mathbf{x}_2, x_3) \in BH\,(0, r_0) . : -r^* \le x_1 \le 0 \right\}.
\end{aligned}
$$

We need the following result analogous to Corollary 53 in two dimensions

LEMMA 74. *For the sets B_+, B^+, B_-, and B^- defined above, we have*

$$
|B^-| = |B_-| \approx |B_+| = |B^+| \approx |B|.
$$

PROOF. The equalities are obvious from the symmetry, so we only need to show the approximate equalities. For convenience we will work with the right half of the ball and first prove that $|B_+| \approx |B|$. We have

$$
\begin{aligned}
|B_+| &= \int\limits_{|\mathbf{y}_2| \le r} \left| B_{2D}\left((0, \mathbf{0}, 0), \sqrt{r^2 - |\mathbf{y}_2|^2} \right) \cap \{ y_1 > r^* \} \right| d\mathbf{y}_2 \\
&\approx \int\limits_{|\mathbf{y}_2| \le r} \left(\sqrt{r^2 - |\mathbf{y}_2|^2} - r^* \right)_+ \cdot h\left(\sqrt{r^2 - |\mathbf{y}_2|^2} \right) d\mathbf{y}_2 \\
&\approx \int\limits_{|\mathbf{y}_2| \le r} \left(\sqrt{r^2 - |\mathbf{y}_2|^2} - r^* \right)_+ \cdot \frac{f\left(\sqrt{r^2 - |\mathbf{y}_2|^2} \right)}{\left| F'\left(\sqrt{r^2 - |\mathbf{y}_2|^2} \right) \right|} d\mathbf{y}_2 \\
&= \int\limits_{|\mathbf{y}_2|^2 \le r^2 - r^{*2}} \left(\sqrt{r^2 - |\mathbf{y}_2|^2} - r^* \right) \cdot \frac{f\left(\sqrt{r^2 - |\mathbf{y}_2|^2} \right)}{\left| F'\left(\sqrt{r^2 - |\mathbf{y}_2|^2} \right) \right|} d\mathbf{y}_2 \\
&\approx \frac{f(r)}{|F'(r)|} \int\limits_{|\mathbf{y}_2|^2 \le r^2 - r^{*2}} \left(\sqrt{r^2 - |\mathbf{y}_2|^2} - r^* \right) d\mathbf{y}_2
\end{aligned}
$$

where for the last equality we used (7.12). Since we trivially have an upper bound $|B_+| \leq |B|$, we only need to obtain the lower bound. Passing to the polar coordinates, $\rho = |\mathbf{y}_2|$, we have

$$\begin{aligned} \int_{|\mathbf{y}_2|^2 \leq r^2 - r^{*2}} \left(\sqrt{r^2 - |\mathbf{y}_2|^2} - r^*\right) d\mathbf{y}_2 &\approx \int_0^{\sqrt{r^2 - r^{*2}}} \left(\sqrt{r^2 - \rho^2} - r^*\right) \rho^{n-3} d\rho \\ &\geq \int_0^{\sqrt{r^2 - \frac{1}{4}(r+r^*)^2}} \left(\sqrt{r^2 - \rho^2} - r^*\right) \rho^{n-3} d\rho \\ &\geq \frac{1}{2}(r - r^*) \int_0^{\sqrt{r^2 - \frac{1}{4}(r+r^*)^2}} \rho^{n-3} d\rho \\ &\approx (r - r^*)^{\frac{n}{2}} r^{\frac{n}{2}-1}. \end{aligned}$$

Using part (2) of Proposition 47 we have $r - r^* \approx 1/|F'(r)|$ and therefore

$$|B_+| \gtrsim \frac{f(r)}{|F'(r)|^{\frac{n}{2}+1}} r^{\frac{n}{2}-1}.$$

Combining with the upper bound and Lemma 57 we conclude

$$|B_+| \approx |B|.$$

Now, for the lower bound $|B_-| \gtrsim |B|$ we note that $B(0, r^*) \subset B_-(0, r) \cup B^-(0, r)$ and by Lemma 57

$$|B(0, r^*)| \approx \frac{f(r^*)}{|F'(r^*)|^{\frac{n}{2}+1}} r^{*\frac{n}{2}-1}.$$

Thus the only thing left to show is that $r^* \approx r$ for $x_1 = 0$. Indeed, combining the estimates from lemmas 48 and 49 with $x_1 = 0$ we get

$$f(r^*)(r - r^*) \leq h(r) \leq \frac{f(r^*)}{|F'(r^*)|},$$

and thus

$$r^* + \frac{1}{|F'(r^*)|} \geq r.$$

Using assumption (4) on the geometry F this gives

$$r^* \left(1 + \frac{1}{\varepsilon}\right) \geq r,$$

which together with the trivial bound $r^* \leq r$ concludes the proof. □

We are now ready to prove the n-dimensional $(1, 1)$-Poincaré inequality, which is used to prove the proportional $(1, 1)$-Sobolev inequality, which in turn is a key ingredient in proving continuity of weak solutions.

PROPOSITION 75. *Let the balls $B(0, r)$ and the degenerate gradient ∇_A be as above. There exists a constant C such that the Poincaré Inequality*

$$\int_{B(0,r)} |w(x) - \bar{w}| \, dx \leq Cr \int_{B(0,2r)} |\nabla_A w| dx$$

holds for any Lipschitz function w and sufficiently small $r > 0$. Here $\bar{w}$ is the average defined by

$$\bar{w} = \frac{1}{|B(0,r)|}\int_{B(0,r)} w dx.$$

PROOF. First note that using Lemma 70 and the symmetry of the problem it is sufficient to show

$$\begin{aligned} I_1 + I_2 &\equiv \frac{1}{|B|}\left(\iint_{B_- \times B_+} |w(x) - w(y)|\, dxdy + \iint_{B^+ \times B_+} |w(x) - w(y)|\, dxdy\right) \\ &\le Cr\int_{2B} |\nabla_A w| dx. \end{aligned}$$

To estimate I_2 we connect two points $(x_1, \mathbf{x}_2, x_3) \in B^+$ and $(y_1, \mathbf{y}_2, y_3) \in B_+$, by straight segments connecting the following pairs of points

$$\begin{aligned} &(x_1, \mathbf{x}_2, x_3) \text{ to } (x_1, \mathbf{x}_2, y_3) \\ &(x_1, \mathbf{x}_2, y_3) \text{ to } (x_1, \mathbf{y}_2, y_3) \\ &(x_1, \mathbf{y}_2, y_3) \text{ to } (y_1, \mathbf{y}_2, y_3). \end{aligned}$$

Note that the curve described above lies entirely in in $Q(r)$ defined in Lemma 73. Moreover, since the first part of the path lies entirely in B^+ we have

$$\begin{aligned} &\iint_{B^+ \times B_+} |w(x_1, \mathbf{x}_2, x_3) - w(x_1, \mathbf{x}_2, y_3)|\, dx_1 d\mathbf{x}_2 dx_3 dy_1 d\mathbf{y}_2 dy_3 \\ &= \iint_{B^+ \times B_+} \left|\int_{x_3}^{y_3} w_t(x_1, \mathbf{x}_2, t) dt\right| dx_1 d\mathbf{x}_2 dx_3 dy_1 d\mathbf{y}_2 dy_3 \\ &\le |B_+|\int_{B^+} \frac{h(r)}{f(x_1)} |\nabla_A w(x_1, \mathbf{x}_2, t)| dx_1 d\mathbf{x}_2 dt \le Cr|B|\int_B |\nabla_A w| dx, \end{aligned}$$

where just as in the 2-dimensional case we used

$$\frac{h(r)}{f(x_1)} \approx \frac{rf(r)}{f(r)} = Cr,$$

since $-r \le x_1 \le -r^*$ in B_+. For the other two parts of the path, the estimates are the same as in the proof of the classical $(1, 1)$ Poincaré in dimensions $n - 2$ and 1 respectively, so we obtain

$$\begin{aligned} &\iint_{B^+ \times B_+} |w(x_1, \mathbf{x}_2, y_3) - w(x_1, \mathbf{y}_2, y_3)|\, dx_1 d\mathbf{x}_2 dx_3 dy_1 d\mathbf{y}_2 dy_3 \\ &\qquad \le Cr|B|\int_{2B} |\nabla_{\mathbf{x}_2} w| dx \le Cr|B|\int_{2B} |\nabla_A w| dx \\ &\iint_{B^+ \times B_+} |w(x_1, \mathbf{y}_2, y_3) - w(y_1, \mathbf{y}_2, y_3)|\, dx_1 d\mathbf{x}_2 dx_3 dy_1 d\mathbf{y}_2 dy_3 \\ &\qquad \le Cr|B|\int_{2B} |\partial_{x_1} w| dx \le Cr|B|\int_{2B} |\nabla_A w| dx. \end{aligned}$$

This concludes the estimate for I_2. To estimate I_1 we similarly connect points in B_- and B_+ by first moving in the first and second variables to reach the set B_+ and then going "vertically" to connect x_3 and y_3. The proof is similar to the 2-dimensional case and is left to the reader. □

3. Proportional vanishing L^1-Sobolev inequality

Our geometric continuity theorem requires the proportional vanishing L^1-Sobolev inequality (2.2), and we now turn to establishing this inequality. For simplicity we consider first the 2-dimensional case.

Define

$$K_r(x,y) \equiv \frac{\widehat{d}(x,y)}{|B(x,d(x,y))|}\mathbf{1}_{\Gamma(x,r)}(y), \tag{8.32}$$

and

$$\Gamma(x,r) = \{y \in B(x,r) : x_1 \leq y_1 \leq x_1 + r,\ |y_2 - x_2| < h^*(x_1, y_1 - x_1)\},$$

and for $y \in \Gamma(x,r)$ let $h_{x,y} = h^*(x_1, y_1 - x_1)$. First, recall from Proposition 51 that we have an estimate

$$|B(x,d(x,y))| \approx h_{x,y}(d(x,y) - d^*(x,y))$$

and by Corollary 59 we have $|B(x,d(x,y))| \approx h_{x,y}\widehat{d}(x,y)$. Thus,

$$K_r(x,y) \approx \frac{1}{h_{x,y}}\mathbf{1}_{\{(x,y):x_1 \leq y_1 \leq x_1 + r,\ |y_2 - x_2| < h_{x,y}\}}(x,y).$$

Now denote the dual cone $\Gamma^*(y,r)$ by

$$\Gamma^*(y,r) \equiv \{x \in B(y,r) : y \in \Gamma(x,r)\}.$$

Then we have

$$\begin{aligned} \Gamma^*(y,r) &= \{x \in B(y,r) : x_1 \leq y_1 \leq x_1 + r,\ |y_2 - x_2| < h_{x,y}\} \\ &= \{x \in B(y,r) : y_1 - r \leq x_1 \leq y_1,\ |x_2 - y_2| < h_{x,y}\}, \end{aligned} \tag{8.33}$$

and consequently we get the 'straight across' estimate in $n = 2$ dimensions,

$$\int K_r(x,y)\,dx \approx \int_{y_1 - r}^{y_1}\left\{\int_{y_2 - h_{x,y}}^{y_2 + h_{x,y}} \frac{1}{h_{x,y}}dx_2\right\}dx_1 \approx \int_{x_1}^{x_1 + r} dy_1 = r\ . \tag{8.34}$$

Turning now to the case of $n \geq 3$ dimensions, we have from (8.21) that

$$K_{B(0,r_0)}(x,y) \approx \begin{cases} \frac{1}{r^{n-1}f(x_1)}\mathbf{1}_{\widetilde{\Gamma}(x,r_0)}(y), & 0 < r = y_1 - x_1 < \frac{2}{|F'(x_1)|} \\ \frac{|F'(x_1+r)|^{n-1}}{f(x_1+r)\lambda(x_1,r)^{n-2}}\mathbf{1}_{\widetilde{\Gamma}(x,r_0)}(y), & R \geq r = y_1 - x_1 \geq \frac{2}{|F'(x_1)|} \end{cases},$$

where $\lambda(x_1, r) \equiv \sqrt{r|F'(x_1 + r)|}$. We denote the size of the kernel $K_{B(0,r_0)}(x,y)$ as $\frac{1}{s_{y_1 - x_1}}$ where

$$\frac{1}{s_{y_1 - x_1}} \equiv \begin{cases} \frac{1}{r^{n-1}f(x_1)}, & 0 < r = y_1 - x_1 < \frac{2}{|F'(x_1)|} \\ \frac{|F'(x_1+r)|^{n}}{f(x_1+r)\lambda(x_1,r)^{n-2}}, & 0 < r = y_1 - x_1 \geq \frac{2}{|F'(x_1)|} \end{cases},$$

and where the quantity s_r can be, roughly speaking, thought of a cross sectional volume analogous to the height h_r in the two dimensional case. We have

$$\begin{aligned}
&\int_{B_+(0,r_0)} K_{B(0,r_0)}(x,y)\,dy \\
&= \sum_{k=0}^{\infty} \int_{x_1+r_{k+1}}^{x_1+r_k} \left[\int_{|\mathbf{x}_2-\mathbf{y}_2|\le\sqrt{r_k^2-r_{k+1}^2}} \left[\int_{x_3-h^*(x_1,r_k)}^{x_3+h^*(x_1,r_k)} \frac{1}{s_{y_1-x_1}} |B(0,r_0)|\,dy_3 \right] d\mathbf{y}_2 \right] dy_1 \\
&= \sum_{k=0}^{\infty} \int_{x_1+r_{k+1}}^{x_1+r_k} \left[\left(\sqrt{r_k^2-r_{k+1}^2}\right)^{n-2} 2h^*(x_1,r_k) \right] \frac{1}{s_{y_1-x_1}} dy_1 \\
&\approx \sum_{k=0}^{\infty} \int_{x_1+r_{k+1}}^{x_1+r_k} s_{y_1-x_1} \frac{1}{s_{y_1-x_1}} dy_1 = \int_{x_1}^{x_1+r_0} dy_1 = r_0,
\end{aligned}$$

where the approximation in the fourth line above comes from the estimates

$$\begin{aligned}
\left(\sqrt{r_k^2-r_{k+1}^2}\right)^{n-2} 2h^*(x_1,r_k)(r_k-r_{k+1}) = \left|\widetilde{E}(x,r_k)\right| &\approx |B(x,r_k)| \\
&\approx s_{r_k}(r_k-r_{k+1}),
\end{aligned}$$

and

$$s_{r_k} \approx s_{y_1-x_1}, \qquad \text{for } x_1+r_{k+1} \le y_1 < x_1+r_k\ .$$

This gives the n-dimensional 'straight across' estimate,

$$\int_{B(0,r_0)} K_{B(0,r_0)}(x,y)\,dy \approx r_0\ . \tag{8.35}$$

We can now prove the proportional vanishing L^1-Sobolev inequality (2.2) by appealing to the the $(1,1)$ *Poincaré* inequality in Proposition 72 from the previous section.

PROPOSITION 76. *Let the balls $B(0,r)$ and the degenerate gradient ∇_A be as above. There exists a constant C such that the proportional vanishing L^1-Sobolev inequality*

$$\int_{B(0,r)} |w|\,dx \le Cr \int_{B(0,2r)} |\nabla_A w| dx, \tag{8.36}$$

holds for any Lipschitz function w that vanishes on a subset E of the ball $B(0,r)$ with $|E| \ge \frac{1}{2}|B(0,r)|$, and all sufficiently small $r>0$.

PROOF. We have

$$\begin{aligned}
\int_{B(0,r)} |w|\,dx &= \int_{B(0,r)} \left| w(x) - \frac{1}{|E\cap B|}\int_{E\cap B} w(y)dy \right| dx \\
&\le \frac{1}{|E\cap B|} \int\int_{B\times E\cap B} |w(x)-w(y)|\,dxdy \\
&\lesssim \frac{1}{|B|} \int\int_{B\times B} |w(x)-w(y)|\,dxdy.
\end{aligned} \tag{8.37}$$

Next, it follows from the proofs of Propositions 72 and 75 that

$$\frac{1}{|B|}\int_{B\times B} |w(x)-w(y)|\,dxdy \le Cr\int_{2B} |\nabla_A w|\,. \tag{8.38}$$

Estimate 8.36 follows from (8.37) and (8.38). □

4. Orlicz Sobolev inequalities for bump functions Φ_N

Recall that the relevant bump functions Φ_N are given in (4.4) by

$$\Phi_N(t) \equiv \begin{cases} t(\ln t)^N & \text{if} \quad t \geq E = E_N = e^{2N} \\ (\ln E)^N t & \text{if} \quad 0 \leq t \leq E = E_N = e^{2N} \end{cases}.$$

Next, define the positive operator $T_{B(0,r_0)} : L^1(\mu_{r_0}) \to L^\Phi(\mu_{r_0})$ by

$$T_{B(0,r_0)} g(x) \equiv \int_{B(0,r_0)} K_{B(0,r_0)}(x,y)\, g(y) dy$$

where $d\mu_{r_0} \equiv \frac{dxdy}{|B(0,r_0)|}$, with kernel $K_{B(0,r_0)}$ defined by

$$K_{B(0,r_0)}(x,y) = \frac{\widehat{d}(x,y)}{|B(x,d(x,y))|} \mathbf{1}_{\widetilde{\Gamma}(r_0)}(x,y),$$

where $\widetilde{\Gamma}$ is given by (8.20), and

$$\widehat{d}(x,y) \equiv \min\left\{ d(x,y), \frac{1}{|F'(x_1 + d(x,y))|} \right\}. \tag{8.39}$$

We will prove the strong form of the norm inequality

$$\left\| T_{B(0,r_0)} g \right\|_{L^\Phi(\mu_{r_0})} \leq C\varphi(r_0)\, \|g\|_{L^1(\mu_{r_0})}, \tag{8.40}$$

which in turn implies the norm inequality

$$\|w\|_{L^\Phi(\mu_{r_0})} \leq C\varphi(r_0)\, \|\nabla_A w\|_{L^1(\mu_{r_0})}, \qquad w \in \left(W_A^{1,1}\right)_0 (B(0,r_0)) \tag{8.41}$$

by the subrepresentation inequality from Lemma 64 with $g = \nabla_A w$. Indeed, we even have a version of (8.41) for each of the half balls $HB(0,r_0) = H_{\text{right}} B(0,r_0)$ and $H_{\text{left}} B(0,r_0)$.

DEFINITION 77. Define $\left(W_A^{1,1}\right)_0 (HB(0,r_0))$ to be the $W_A^{1,1}$-closure of those Lipschitz functions w in $HB(0,r_0)$ that vanish in a Euclidean neighbourhood of the compact set $\{x \in \partial B(0,r_0) : x_1 \geq 0\}$, and similarly for $\left(W_A^{1,1}\right)_0 (H_{\text{left}} B(0,r_0))$.

LEMMA 78. *Assume that the strong form of the norm inequality* (8.40) *holds. Then the standard form of the norm inequality* (8.41) *holds, and moreover, we also have the halfball inequalities*

$$\|w\|_{L^\Phi\left(HB_{\text{right/left}}(0,r_0),\mu_{r_0}\right)} \leq C\varphi(r_0)\, \|\nabla_A w\|_{L^1\left(HB_{\text{right/left}}(0,r_0),\mu_{r_0}\right)}, \tag{8.42}$$

$w \in \left(W_A^{1,1}\right)_0 \left(HB_{\text{right/left}}(0,r_0)\right)$, *where we take the same choice of* $H_{\text{right}} B(0,r_0)$ *or* $H_{\text{left}} B(0,r_0)$ *on both sides of the inequality* (8.42).

PROOF. Given a radius $r_0 > 0$, choose $r_{-1} > r_0$ so that $r_0 = (r_{-1})^*$. Then extend $w \in W_0^{1,1}(B(0,r_0))$ to $w \in W_0^{1,1}(B(0,r_{-1}))$ by defining w to vanish outside $B(0,r_0)$. Now for each x in the smaller halfball $HB(0,r_0)$, we have

$$E(x,s) \cap HB(0,r_{-1}) \subset E(0,r_{-1}) \text{ and } |E(x,s) \cap HB(0,r_{-1})| \approx |E(0,r_{-1})|$$

for an appropriate $s > 0$. Now we apply Lemma 64 in the larger halfball $HB(0,r_{-1})$, together with the fact that w vanishes on the end $E(x,s)$, to conclude that

$$|w(x)| = |w(x) - E(x,s)| \lesssim T_{B(0,r_{-1})} \left(\mathbf{1}_{HB(0,r_{-1})} |\nabla_A w| \right)(x),$$

for $x \in HB(0, r_{-1})$. Thus from (8.40) we obtain

$$\begin{aligned} \|w\|_{L^\Phi(HB(0,r_{-1}),\mu_{r_{-1}})} &\leq \left\|T_{B(0,r_{-1})}\left(\mathbf{1}_{HB(0,r_{-1})}|\nabla_A w|\right)\right\|_{L^\Phi(HB(0,r_{-1}),\mu_{r_{-1}})} \\ &\leq C\varphi(r_{-1}) \ \||\nabla_A w|\|_{L^1(HB(0,r_{-1}),\mu_{r_{-1}})} . \end{aligned}$$

This gives the case of 8.41 for the halfball $HB(0, r_0)$ upon noting that both w and $|\nabla_A w|$ vanish outside $B(0, r_0)$, and $r_0 = (r_{-1})^*$, and so easy estimates show that we actually have $\varphi(r_{-1}) \approx \varphi(r_0)$ and both

$$\begin{aligned} \|w\|_{L^\Phi(HB(0,r_{-1}),\mu_{r_{-1}})} &\approx \|w\|_{L^\Phi(HB(0,r_0),\mu_{r_0})} , \\ \||\nabla_A w|\|_{L^1(HB(0,r_{-1}),\mu_{r_{-1}})} &\approx \||\nabla_A w|\|_{L^1(HB(0,r_0),\mu_{r_0})} . \end{aligned}$$

□

We begin by proving that the bound (8.40) holds if the following endpoint inequality holds:

$$\Phi^{-1}\left(\sup_{y\in B}\int_B \Phi\left(K(x,y)|B|\alpha\right)d\mu(x)\right) \leq C\alpha\varphi(r) . \tag{8.43}$$

for all $\alpha > 0$.

LEMMA 79. *The endpoint inequality (8.43) implies the norm inequality (8.41).*

PROOF. If (8.43) holds, then Jensen's inequality applied to the convex function Φ gives

$$\begin{aligned} &\int_B \Phi\left(\frac{T_{B(0,r_0)}g}{C_1\varphi(r_0)\ \|g\|_{L^1(\mu)}}\right)d\mu(x) \\ &\lesssim \int_B \Phi\left(\int_B K(x,y)\ |B|\ \frac{1}{C_1\varphi(r_0)}\ \frac{g(y)\,d\mu(y)}{\|g\|_{L^1(\mu)}}\right)d\mu(x) \\ &\leq \int_B\int_B \Phi\left(K(x,y)\ |B|\ \frac{1}{C_1\varphi(r_0)}\right)\frac{g(y)\,d\mu(y)}{\|g\|_{L^1(\mu)}}d\mu(x) \\ &\leq \int_B\left\{\sup_{y\in B}\int_B \Phi\left(K(x,y)\ |B|\ \frac{1}{C_1\varphi(r_0)}\right)d\mu(x)\right\}\frac{g(y)\,d\mu(y)}{\|g\|_{L^1(\mu)}} \\ &\leq \Phi\left(\frac{C}{C_1}\right)\int_B \frac{g(y)\,d\mu(y)}{\|g\|_{L^1(\mu)}} \leq 1, \end{aligned}$$

for C_1 sufficiently large, and where we used (8.43) with $\alpha = \frac{1}{C_1\varphi(r_0)} \equiv \frac{\Phi^{-1}(1)}{C\varphi(r_0)}$. We conclude from the definition of $L^\Phi(\mu_{r_0})$ that

$$\left\|T_{B(0,r_0)}g\right\|_{L^\Phi(\mu_{r_0})} \leq C_1\varphi(r_0)\ \|g\|_{L^1(\mu_{r_0})} .$$

□

PROPOSITION 80. *Let $n \geq 2$. Assume that for some $C > 0$ the function*

$$\varphi(r) \equiv C|F'(r)|^N r^{N+1} \tag{8.44}$$

satisfies $\lim_{r\to 0}\varphi(r) = 0$. *Assume in addition that geometry F satisfies*

$$F''(r) \leq \left(1+\frac{1-\varepsilon}{N}\right)\frac{|F'(r)|}{r}, \quad r\in(0,r_0), \quad \varepsilon>0. \tag{8.45}$$

Then:

(1) *the* (Φ,φ)*-Sobolev inequality* (8.40) *holds with geometry* F*, with* φ *as in* (8.44)*, and with* Φ *as in* (4.4)*,* $N>1$*,*
(2) *and if* $\varphi_{\max}(r)\equiv\sup_{0<s<r_0}\varphi(s)<\infty$ *is a finite constant function, then the* $(\Phi,\varphi_{\max})$*-Sobolev inequality* (8.40) *holds with geometry* F*, with* φ *as in* (8.44)*, and with* Φ *as in* (4.4)*,* $N>1$*,*
(3) *in particular, if for some* $\varepsilon>0$ *we have*

$$|F'(r)|\leq C\left(\frac{1}{r}\right)^{1+\frac{1-\varepsilon}{N}}, \tag{8.46}$$

then the $(\Phi,\varphi_{\max})$*-Sobolev inequality* (8.40) *holds with geometry* F *and* $\varphi_{\max}(r)\equiv C$*.*

Proof. For Part (1) it suffices to prove the endpoint inequality

$$\Phi^{-1}\left(\sup_{y\in B}\int_B\Phi\left(K(x,y)|B|\alpha\right)d\mu(x)\right)\leq C\alpha\varphi(r(B)),\qquad\alpha>0. \tag{8.47}$$

for the balls and kernel associated with our geometry F, the Orlicz bump Φ, and the function $\varphi(r)$ satisfying (8.44). Fix parameters $N>1$ and $t_N>1$. Now we consider the specific function $\omega(r(B))$ given by

$$\omega(r(B))=\frac{1}{t_N|F'(r(B))|}.$$

Using the submultiplicativity of Φ we have

$$\begin{aligned}\int_B\Phi\left(K(x,y)|B|\alpha\right)d\mu(x) &= \int_B\Phi\left(\frac{K(x,y)|B|}{\omega(r(B))}\alpha\omega(r(B))\right)d\mu(x)\\ &\leq \Phi\left(\alpha\omega(r(B))\right)\int_B\Phi\left(\frac{K(x,y)|B|}{\omega(r(B))}\right)d\mu(x)\end{aligned}$$

and we will now prove

$$\int_B\Phi\left(\frac{K(x,y)|B|}{\omega(r(B))}\right)d\mu(x)\leq C_N\varphi(r(B))|F'(r(B))|, \tag{8.48}$$

for all small balls B of radius $r(B)$ centered at the origin. Altogether this will give us

$$\int_B\Phi\left(K(x,y)|B|\alpha\right)d\mu(x)\leq C_N\varphi(r(B))|F'(r(B))|\,\Phi\left(\frac{\alpha}{t_N|F'(r(B))|}\right).$$

Now we note that $x\Phi(y)=xy\frac{\Phi(y)}{y}\leq xy\frac{\Phi(xy)}{xy}=\Phi(xy)$ for $x\geq1$ since $\frac{\Phi(t)}{t}$ is monotone increasing. But from (8.44) and assumptions (1) and (4) on the geometry F we have $\varphi(r)|F'(r)|\gg1$ and so

$$\begin{aligned}\int_B\Phi\left(K(x,y)|B|\alpha\right)d\mu(x) &\leq \Phi\left(C_N\varphi(r(B))|F'(r(B))|\,\alpha\frac{1}{t_N|F'(r(B))|}\right)\\ &= \Phi\left(\frac{C_N}{t_N}\alpha\varphi(r(B))\right),\end{aligned}$$

which is (8.47) with $C=\frac{C_N}{t_N}$. Thus it remains to prove (8.48).

So we now take $B=B(0,r_0)$ with $r_0\ll1$ so that $\omega(r(B))=\omega(r_0)$. First, recall

$$|B(0,r_0)|\approx\frac{f(r_0)}{|F'(r_0)|^n}\lambda(0,r_0)^{n-2},$$

where we recall from (8.22) that

$$\lambda(x_1,r) \equiv \sqrt{r\left|F'(x_1+r)\right|},$$

and we now denote the size of the kernel $K_{B(0,r_0)}(x,y)$ as $\frac{1}{s_{y_1-x_1}}$ where

$$\frac{1}{s_{y_1-x_1}} \equiv \begin{cases} \frac{1}{r^{n-1}f(x_1)}, & 0<r=y_1-x_1<\frac{2}{|F'(x_1)|} \\ \frac{|F'(y_1)|^n}{f(y_1)(r|F'(y_1)|)^{\frac{n-2}{2}}}, & 0<r=y_1-x_1\geq\frac{2}{|F'(x_1)|} \end{cases}.$$

We are writing the size of $\frac{1}{K_{B(0,r_0)}(x,y)}$ as $s_{y_1-x_1}=s_r$ since the quantity s_r can be, roughly speaking, thought of a cross sectional volume analogous to the height h_r in the two dimensional case.

Next, write $\Phi(t)$ as

$$\Phi(t)=t\Psi(t), \qquad \text{for } t>0, \tag{8.49}$$

where for $t\geq E$,

$$\begin{aligned} t\Psi(t) &= \Phi(t)=t(\ln t)^N \\ &\Longrightarrow \Psi(t)=(\ln t)^N, \end{aligned}$$

and for $t<E$,

$$\begin{aligned} t\Psi(t) &= \Phi(t)=t(\ln E)^N \\ &\Longrightarrow \Psi(t)=(\ln E)^N. \end{aligned}$$

Now temporarily fix $y=(y_1,\mathbf{y}_2,y_3)\in B_+(0,r_0)\equiv\{x\in B(0,r_0):x_1>0\}$. Using the definition of $\widetilde{\Gamma}(x,r_0)$ in (8.20) above, we have for $0<a<b<r_0$ that

$$\begin{aligned}
\mathcal{I}_{a,b}(y) &\equiv \int_{\{x\in B_+(0,r_0):a\leq y_1-x_1\leq b\}\cap\widetilde{\Gamma}^*(y,r_0)} \Phi\left(K_{B(0,r_0)}(x,y)\frac{|B(0,r_0)|}{\omega(r_0)}\right)\frac{dx}{|B(0,r_0)|} \\
&= \sum_{k:\ a\leq r_{k+1}<r_k\leq b}\int_{y_1-r_k}^{y_1-r_{k+1}} \\
&\quad\times\left[\int_{|\mathbf{x}_2-\mathbf{y}_2|\leq\sqrt{r_k^2-r_{k+1}^2}}\left[\int_{y_3-h^*(x_1,r_k)}^{y_3+h^*(x_1,r_k)}\Phi\left(\frac{1}{s_{y_1-x_1}}\frac{|B(0,r_0)|}{\omega(r_0)}\right)dx_3\right]d\mathbf{x}_2\right]\frac{dx_1}{|B(0,r_0)|} \\
&= \sum_{k:\ a\leq r_{k+1}<r_k\leq b}\int_{y_1-r_k}^{y_1-r_{k+1}} \\
&\quad\times\left[4\left(\sqrt{r_k^2-r_{k+1}^2}\right)^{n-2}h^*(x_1,r_k)\right]\Phi\left(\frac{1}{s_{y_1-x_1}}\frac{|B(0,r_0)|}{\omega(r_0)}\right)\frac{dx_1}{|B(0,r_0)|} \\
&\approx \int_{y_1-b}^{y_1-a} s_{y_1-x_1}\Phi\left(\frac{1}{s_{y_1-x_1}}\frac{|B(0,r_0)|}{\omega(r_0)}\right)\frac{dx_1}{|B(0,r_0)|} \\
&= \int_{y_1-b}^{y_1-a} s_{y_1-x_1}\left(\frac{1}{s_{y_1-x_1}}\frac{|B(0,r_0)|}{\omega(r_0)}\right)\Psi\left(\frac{1}{s_{y_1-x_1}}\frac{|B(0,r_0)|}{\omega(r_0)}\right)\frac{dx_1}{|B(0,r_0)|},
\end{aligned}$$

where the approximation in the fourth line above comes from the estimates from Lemma 57 and Lemma 60

$$\begin{aligned} \left(\sqrt{r_k^2-r_{k+1}^2}\right)^{n-2}h^*(x_1,r_k)(r_k-r_{k+1}) &= \left|\widetilde{E}(x,r_k)\right|\approx|B(x,r_k)| \\ &\approx s_{r_k}(r_k-r_{k+1}), \end{aligned}$$

and

$$s_{r_k} \approx s_{y_1-x_1}, \qquad \text{for } r_{k+1} \le y_1 - x_1 < r_k \ .$$

Thus we have

$$\begin{aligned} \mathcal{I}_{a,b}(y) &\approx \frac{1}{\omega(r_0)} \int_{y_1-b}^{y_1-a} \Psi\left(\frac{1}{s_{y_1-x_1}} \frac{|B(0,r_0)|}{\omega(r_0)}\right) dx_1 \\ &= \frac{1}{\omega(r_0)} \int_a^b \Psi\left(\frac{1}{s_{y_1-x_1}} \frac{|B(0,r_0)|}{\omega(r_0)}\right) dr. \end{aligned}$$

and so

$$\begin{aligned} &\int_{B_+(0,r_0)} \Phi\left(K_{B(0,r_0)}(x,y) \frac{|B(0,r_0)|}{\omega(r_0)}\right) \frac{dx}{|B(0,r_0)|} \\ &= \mathcal{I}_{0,y_1}(y) \\ &= \frac{2}{\omega(r_0)} \int_0^{y_1} \Psi\left(\frac{1}{s_r} \frac{|B(0,r_0)|}{\omega(r_0)}\right) dr \ . \end{aligned}$$

It thus suffices to prove

$$\mathcal{I}_{0,y_1} = \frac{1}{\omega(r_0)} \int_0^{y_1} \Psi\left(\frac{1}{s_r} \frac{|B(0,r_0)|}{\omega(r_0)}\right) dr \le C_N \varphi(r_0) |F'(r_0)| \ , \tag{8.50}$$

where $|B(0,r_0)|$ is now the Lebesgue measure of the n-dimensional ball $B(0,r_0) = B_{nD}(0,r_0)$.

To prove this we divide the interval $(0,y_1)$ of integration in r into three regions as before:

(1): the small region $\mathcal{S}$ where $\frac{|B(0,r_0)|}{s_r \varphi(r_0)} \le E$,

(2): the big region $\mathcal{R}_1$ that is disjoint from $\mathcal{S}$ and where $r = y_1 - x_1 < \frac{2}{|F'(x_1)|}$ and

(3): the big region $\mathcal{R}_2$ that is disjoint from $\mathcal{S}$ and where $r = y_1 - x_1 \ge \frac{2}{|F'(x_1)|}$.

In the small region $\mathcal{S}$ we use that Φ is linear on $[0,E]$ to obtain that the integral in the right hand side of (8.50), when restricted to those $r \in (0,r_0)$ for which $\frac{|B(0,r_0)|}{s_r \omega(r_0)} \le E$, is bounded by

$$\begin{aligned} &\frac{1}{\omega(r_0)} \int_0^{r_0} \Psi\left(\frac{|B(0,r_0)|}{s_r \omega(r_0)}\right) dr \\ &= \frac{1}{\omega(r_0)} \int_0^{r_0} \frac{\Phi(E)}{E} dr = \frac{1}{\omega(r_0)} (\ln E)^N r_0 \\ &\le C\, t_N\, r_0 |F'(r_0)| \le C_N \varphi(r_0) |F'(r_0)| \ , \end{aligned}$$

since $\omega(r_0) = \frac{1}{t_N |F'(r_0)|}$, and for the last inequality we used $r_0 \le \varphi(r_0)$ which follows from (8.44) and assumtion (1) on the geometry.

We now turn to the first big region $\mathcal{R}_1$ where we have $s_{y_1-x_1} \approx r^{n-1} f(x_1)$. We have using the definition of Ψ (8.49)

$$\int_{\mathcal{R}_1} \Phi\left(K_{B(0,r_0)}(x,y) \frac{|B(0,r_0)|}{\omega(r_0)}\right) \frac{dy}{|B(0,r_0)|} \lesssim \frac{1}{\omega(r_0)} \int_0^{y_1} \ln\left(\frac{c(r_0)}{f(y_1-r) r^{n-1}}\right)^N dr,$$

where we have used the notation

$$c(r_0) \equiv \frac{|B(0,r_0)|}{\omega(r_0)} \approx \frac{f(r_0) r_0^{\frac{n}{2}-1}}{|F'(r_0)|^{\frac{n}{2}}} \lesssim f(r_0) r_0^{n-1}, \tag{8.51}$$

where we used Lemma 57 and for the last inequality, property (4) of geometry F. Using this, we have

$$\int_0^{y_1} \ln\left(\frac{c(r_0)}{f(y_1-r)r^{n-1}}\right)^N dr \le \int_0^{r_0} \ln\left(\frac{f(r_0)r_0^{n-1}}{f(y_1-r)r^{n-1}}\right)$$
$$\le Cr_0 + \int_0^{r_0} (F(x_1)-F(r_0))^N dx_1 + \int_0^{r_0} \left(\ln\frac{r_0}{r}\right)^N dr.$$

For the second integral we have

$$\int_0^{r_0} \left(\ln\frac{r_0}{r}\right)^N dr \lesssim r\left(\ln\frac{r_0}{r}\right)^N \Big|_{r=0}^{r_0} = Cr_0,$$

which can be absorbed into the first term. To estimate the first integral we write

$$I \equiv \int_0^{r_0} (F(x_1)-F(r_0))^N dx_1 = \int_0^{r_0}\left(\int_{x_1}^{r_0} -F'(s)ds\right)^N dx_1$$
$$= \int_0^{r_0}\left(\int_0^{r_0} \chi_{[x_1,r_0]}(s)(-F'(s))ds\right)^N dx_1$$

and we will use Minkowski integral inequality to estimate this. More precisely, we have

$$\left(\int_0^{r_0}\left(\int_0^{r_0} \chi_{[x_1,r_0]}(s)(-F'(s))ds\right)^N dx_1\right)^{\frac{1}{N}}$$
$$\le \int_0^{r_0}\left(\int_0^{r_0} \left(\chi_{[x_1,r_0]}(s)(-F'(s))\right)^N dx_1\right)^{\frac{1}{N}} ds$$
$$= \int_0^{r_0} (-F'(s))\left(\int_0^{r_0} \chi_{[x_1,r_0]}(s)dx_1\right)^{\frac{1}{N}} ds$$
$$= \int_0^{r_0} |F'(s)|s^{\frac{1}{N}} ds.$$

Finally, integrating by parts and using (8.45) we obtain

$$\int_0^{r_0} |F'(s)|s^{\frac{1}{N}} ds = |F'(s)|s^{\frac{1}{N}+1}\frac{1}{\frac{1}{N}+1}\Big|_{r=0}^{r_0} + \frac{1}{\frac{1}{N}+1}\int F''(s)s^{\frac{1}{N}+1}ds$$
$$\left(\frac{1}{N}+1\right)\int_0^{r_0} |F'(s)|s^{\frac{1}{N}} ds = |F'(s)|s^{\frac{1}{N}+1}\Big|_{r=0}^{r_0} + \int F''(s)s^{\frac{1}{N}+1}ds$$
$$\le |F'(s)|s^{\frac{1}{N}+1}\Big|_{r=0}^{r_0} + \left(\frac{1-\varepsilon}{N}+1\right)\int_0^{r_0} |F'(s)|s^{\frac{1}{N}} ds$$
$$\int_0^{r_0} |F'(s)|s^{\frac{1}{N}} ds \le \frac{N}{\varepsilon}|F'(s)|s^{\frac{1}{N}+1}\Big|_{r=0}^{r_0} = \frac{N}{\varepsilon}|F'(r_0)|r_0^{\frac{1}{N}+1},$$

where in the last inequality we used (8.44). This implies

$$I \le C|F'(r_0)|^N r_0^{N+1},$$

and therefore using $r_0 \le C|F'(r_0)|^N r_0^{N+1}$ which follows from the assumption (4) on the geometry, we conclude (8.50) for region $\mathcal{R}_1$.

We now turn to second region $\mathcal{R}_2$, where $r = y_1 - x_1 \geq \frac{2}{|F'(x_1)|}$, and so $s_{y_1-x_1} \approx \frac{f(y_1)r^{\frac{n}{2}-1}}{|F'(y_1)|^{\frac{n}{2}+1}}$. The integral to be estimated becomes

$$I_{\mathcal{R}_2} \equiv \frac{1}{\omega(r_0)} \int_0^{y_1} \ln\left(\frac{c(r_0)|F'(y_1)|^{\frac{n}{2}+1}}{f(y_1)r^{\frac{n}{2}-1}}\right)^N dr,$$

where $c(r_0)$ as in (8.51). Similar to the estimate for region $\mathcal{R}_1$ we have

$$\begin{aligned}\int_0^{y_1} \ln\left(\frac{c(r_0)|F'(y_1)|^{\frac{n}{2}+1}}{f(y_1)r^{\frac{n}{2}-1}}\right)^N dr \lesssim Cr_0 &+ \int_0^{y_1}\left(\ln\frac{r_0}{r}\right)^N dr\\ &+ \int_0^{y_1}(F(y_1)-F(r_0))^N dr\\ &+ \int_0^{y_1}\left(\ln\frac{|F'(y_1)|}{|F'(r_0)|}\right)^N dr.\end{aligned}$$

The first integral was estimated above and can be bounded by Cr_0. For the second integral we can use the trivial estimate

$$\int_0^{y_1}(F(y_1)-F(r_0))^N dr \leq \int_0^{r_0}(F(r)-F(r_0))^N dr,$$

which reduces it to the integral I arising in region $\mathcal{R}_1$. It remains to estimate the third integral which we denote as

$$\mathcal{F}(y_1) \equiv y_1\left(\ln\frac{|F'(y_1)|}{|F'(r_0)|}\right)^N.$$

First note that from the doubling condition on $|F'|$ which is assumption (3), it follows that $\mathcal{F}(0) = 0$. Also, clearly $\mathcal{F}(r_0) = 0$ and thus $\mathcal{F}(y_1)$ achieves a maximum inside the interval $(0, r_0)$. Differentiating $\mathcal{F}(y_1)$ and setting the derivative equal to zero, we obtain the following implicit expression for y_1^* maximizing $\mathcal{F}(y_1)$:

$$\ln\frac{|F'(y_1^*)|}{|F'(r_0)|} = Ny_1^*\frac{F''(y_1^*)}{|F'(y_1^*)|}.$$

Substituting back into the definition of $\mathcal{F}$ and using assumption (5) on the geometry, we get

$$\mathcal{F}(y_1^*) = y_1^*\left(Ny_1^*\frac{F''(y_1^*)}{|F'(y_1^*)|}\right)^N \leq Cy_1^* \leq Cr_0.$$

Parts (2) and (3) of the theorem follow easily from part (1). □

5. Weak Sobolev inequality

Now we turn to the global Sobolev inequality needed for the maximum principle, namely

$$\|w\|_{L^\Phi(\Omega)} \leq C(\Omega)\|\nabla_A w\|_{L^1(\Omega)}, \qquad w \in \left(W_A^{1,1}\right)_0(\Omega) \tag{8.52}$$

where $\Phi = \Phi_N$ with $N > 1$ as defined in (4.6).

PROPOSITION 81. *Assume that for some $C > 0$ the function*

$$\varphi(r) \equiv C|F'(r)|^N r^{N+1}$$

satisfies $\sup_{0<s<r_0}\varphi(s) < \infty \ \forall r_0 > 0$. *Then the global Sobolev inequality (8.52) holds with geometry F and* $C(\Omega) = \sup_{0<s<d(0,\Omega)+diam(\Omega)}\varphi(s)$.

PROOF. This is a consequence of Proposition 80. First, assume that for r_0 as in Proposition 80 we have $\Omega \subset B(0, r_0)$. Extending w to be 0 outside Ω we may assume $w \in \left(W_A^{1,2}\right)_0 (B(0, r_0))$ and the result follows from part (2) of Proposition 80. More generally, for any bounded domain Ω, we construct a finite partition of unity $\{\eta_k\}_{k=1}^K$ consisting of Lipschitz functions η_k supported in metric balls $B_k \equiv B((x_1^k, \mathbf{x}_2^k, x_3^k), r_0/4)$. By translation invariance of the problem in $(\mathbf{x}_2, x_3)$ variables we may assume $(\mathbf{x}_2^k, x_3^k) = (\mathbf{0}, 0)$. Now if $|x_1^k| < 3r_0/4$ then $B_k \subseteq B(0, r_0)$ and we can apply (8.41) to $w\eta_k$ in $B(0, r_0)$. Otherwise, $|x_1^k| \geq 3r_0/4$, and $\forall x \in B_k$ we have $|x| > r_0/2$ so the matrix A is elliptic there and the Orlicz bump Sobolev inequality follows from the classical Sobolev inequality. Finally writing $w = \sum_{k=1}^K w\eta_k$ we obtain the result.

□

CHAPTER 9

Geometric Theorems

For convenience we recall from the introduction our three geometric theorems dealing with respectively local boundedness, the maximum principle and continuity.

THEOREM 82. *Suppose that $\Omega \subset \mathbb{R}^n$ is a domain in $\mathbb{R}^n$ with $n \geq 2$ and that*

$$\mathcal{L}u \equiv \operatorname{div} \mathcal{A}(x,u) \nabla u, \qquad x = (x_1, ..., x_n) \in \Omega,$$

where $\mathcal{A}(x,z) \sim \begin{bmatrix} I_{n-1} & 0 \\ 0 & f(x_1)^2 \end{bmatrix}$, *$I_{n-1}$ is the $(n-1) \times (n-1)$ identity matrix, $\mathcal{A}$ has bounded measurable components, and the geometry $F = -\ln f$ satisfies the structure conditions in Definition 16.*

(1) *If $F \leq F_\sigma$ for some $0 < \sigma < 1$, then every weak subsolution to $\mathcal{L}u = \phi$ with A-admissible ϕ is locally bounded in Ω.*
(2) *On the other hand, if $n \geq 3$ and $\sigma \geq 1$, then there exists a locally unbounded weak solution u in a neighbourhood of the origin in $\mathbb{R}^n$ to the equation $Lu = 0$ with geometry $F = F_\sigma$.*

THEOREM 83. *Suppose that F satisfies the geometric structure conditions in Definition 16 and $F \leq F_\sigma$ for some $0 < \sigma < 1$. Assume that u is a weak subsolution to $\mathcal{L}u = \phi$ in a domain $\Omega \subset \mathbb{R}^n$ with $n \geq 2$, where $\mathcal{L}$ has degeneracy F and ϕ is A-admissible. Moreover, suppose that u is bounded in the weak sense on the boundary $\partial\Omega$. Then u is globally bounded in Ω and satisfies*

$$\sup_{\Omega} u \leq \sup_{\partial\Omega} u + C \|\phi\|_{X(\Omega)}$$

with the constant C depending only on Ω.

THEOREM 84. *Suppose that $\Omega \subset \mathbb{R}^n$ is a domain in $\mathbb{R}^n$ with $n \geq 2$ and that*

$$\mathcal{L}u \equiv \operatorname{div} \mathcal{A}(x,u) \nabla u, \qquad x = (x_1, ..., x_n) \in \Omega,$$

where $\mathcal{A}(x,z) \sim \begin{bmatrix} I_{n-1} & 0 \\ 0 & f(x_1)^2 \end{bmatrix}$, *$I_{n-1}$ is the $(n-1) \times (n-1)$ identity matrix, $\mathcal{A}$ has bounded measurable components, and the geometry $F = -\ln f$ satisfies the structure conditions in Definition 16.*

(1) *If $F \leq F_{3,\sigma}$ for some $0 < \sigma < 1$, then every weak solution to $\mathcal{L}u = 0$ is continuous in Ω.*
(2) *On the other hand, if $n \geq 3$ and $\sigma \geq 1$, then there exists a locally unbounded weak solution u in a neighbourhood of the origin in $\mathbb{R}^n$ to the equation $Lu = 0$ with geometry $F = F_{0,\sigma}$.*

1. Proofs of sufficiency

The first part of the geometric Theorem 82 follows from the abstract Theorem 10 together with the Orlicz-Sobolev inequality in Proposition 80 and the proportional vanishing L^1-Sobolev inequality in Proposition 76. Indeed, in the special case of the degenerate geometry $F_\sigma = \frac{1}{r^\sigma}$, we have

$$|F_\sigma'(r)| = \frac{\sigma}{r^{\sigma+1}}, \quad F_\sigma''(r) = \frac{\sigma(\sigma+1)}{r^{\sigma+2}},$$

and it is easy to check that the conditions of Proposition 80 are satisfied iff $\sigma < \frac{1}{N}$. Thus the superradius $\varphi(r)$ is given by

$$\varphi(r) = Cr^{-\sigma N+1} < \infty,$$

and so the Inner Ball Inequality (4.7) in Proposition 27 shows that weak subsolutions u to the inhomogeneous equation $\mathcal{L}u = \phi$ are locally bounded above, and hence that solutions u are locally bounded.

Similarly, the geometric Theorem 83 follows from the abstract Theorem 30 and Proposition 81.

The first part of the geometric Theorem 84 follows from the abstract Theorem 21 together with the Inner Ball Inequality in Proposition 27, which uses the Orlicz-Sobolev inequality in Proposition 80 as follows. For the geometry $F_{\sigma,k}$, i.e. $F(r) = F_{\sigma,k} = \left(\ln \frac{1}{r}\right)\left(\ln^{(k)} \frac{1}{r}\right)^\sigma$, we have

$$|F'(r)| \approx \frac{1}{r}\left(\ln^{(k)} \frac{1}{r}\right)^\sigma, \quad F''(r) \leq (1+\varepsilon)\frac{|F'(r)|}{r},$$

where ε can be made arbitrarily small by choosing r_0 sufficiently small. Thus the conditions of Proposition (80) are satisfied and we have the following estimate for the superradius:

$$\varphi(r) \leq Cr\left(\ln^{(k)} \frac{1}{r}\right)^{\sigma N}.$$

Now we use the formula (4.8),

$$A_{N,\varepsilon}(r) \equiv C_1 \exp\left\{C_2\left(\frac{\varphi(r)}{r}\right)^{\frac{1}{N-1-\varepsilon}}\right\},$$

for the constant $A_{N,\varepsilon}(r)$ in the Inner Ball Inequality in Proposition 27, together with $A_{N,\varepsilon}(3r) = \frac{1}{2\sqrt{\delta(r)}}$ in Theorem 21, to obtain that for the geometry $F = F_{k,\sigma}$ we have

$$\delta(r) = \frac{1}{4A_{N,\varepsilon}(3r)^2} = \frac{1}{4C_1^2}e^{-2C_2\left(\frac{\varphi(3r)}{3r}\right)^{\frac{1}{N-1-\varepsilon}}} \geq e^{-C'\left(\ln^{(k)} \frac{1}{r}\right)^{\frac{\sigma N}{N-1-\varepsilon}}}.$$

Note that if $\sigma < 1$ we can find $N > 1$ satisfying

$$N > \frac{1+\varepsilon}{1-\sigma},$$

which implies

$$\gamma \equiv \frac{\sigma N}{N-1-\varepsilon} < 1.$$

Now choose $r_j = 4^{-j} r_0$ and let $k = 3$, we thus have

$$\frac{1}{\delta(r_j)^2} \leq e^{C''\left[\ln^{(3)}\left(\frac{1}{r_j}\right)\right]^\gamma}.$$

Now we use $\lambda_j \equiv \frac{1}{2^{3+\frac{C_3}{\delta^2(r_j)}}}$ from (6.4), together with the inequality

$$\left[\ln^{(k)} a\right]^\gamma \leq \varepsilon \ln^{(k)} a, \qquad \text{for } a \text{ sufficiently large depending on } \gamma < 1 \text{ and } \varepsilon > 0,$$

to obtain

$$\begin{aligned}
\sum_{j=1}^\infty \lambda_j &= \sum_{j=1}^\infty \frac{1}{2^{3+\frac{C_3}{\delta^2(r_j)}}} \approx \sum_{j=1}^\infty \frac{1}{2^{3+C_3 e^{C''\left[\ln^{(3)}\left(\frac{1}{r_j}\right)\right]^\gamma}}} \\
&\gtrsim \sum_{j=1}^\infty 2^{-C'''\varepsilon \ln^{(2)}\left(\frac{4^j}{r_0}\right)} \gtrsim \sum_{j=1}^\infty e^{-C''''\varepsilon \ln j} = \sum_{j=1}^\infty \frac{1}{j^{C''''\varepsilon}} = \infty,
\end{aligned}$$

if $\varepsilon > 0$ is chosen sufficiently small. This establishes the hypotheses of Theorem 21 and completes the proof of the first part of the geometric Theorem 84.

The second parts of the geometric Theorems 82 and 84 follow from the counterexample constructed in the next section.

2. An unbounded weak solution

In this final section of the final chapter of Part 3, we demonstrate that weak solutions to our degenerate equations can fail to be locally bounded. We modify an example of Morimoto [**Mor**] that was used to provide an alternate proof of a result of Kusuoka and Strook [**KuStr**].

THEOREM 85. *Suppose that $g \in C^\infty(\mathbb{R})$ satisfies $g(x) \geq 0$, $g(0) = 0$ and the decay condition*

$$\liminf_{x\to 0} \left| x \ln \frac{1}{g(x)} \right| \neq 0. \tag{9.1}$$

Then for some $\varepsilon > 0$, the operator

$$L \equiv \frac{\partial^2}{\partial x^2} + g(x)\frac{\partial^2}{\partial y^2} + \frac{\partial^2}{\partial t^2}$$

***fails** to be $W_A^{1,2}(\mathbb{R}^2)$-hypoelliptic in an open subset $(-1,1) \times \mathbb{R} \times (-\varepsilon, \varepsilon)$ of $\mathbb{R}^3$ containing the origin, where $\nabla_A \equiv \left(\frac{\partial}{\partial x}, \sqrt{g(x)}\frac{\partial}{\partial y}, \frac{\partial}{\partial t}\right)$ is the degenerate gradient associated with $\mathcal{L}$. In fact, we construct a weak solution u of the homogeneous equation $\mathcal{L}u = 0$ in $(-1,1) \times \mathbb{T} \times (-\varepsilon, \varepsilon)$ with $\|u\|_{L^\infty((-\delta,\delta)\times\mathbb{T}\times(-\varepsilon',\varepsilon'))} = \infty$ for all $\delta > 0$ and $0 < \varepsilon' < \varepsilon$.*

PROOF. For $a, \eta > 0$ we follow Morimoto [**Mor**], who in turn followed Bouendi and Goulaouic [**BoGo**], by considering the second order operator $L_\eta \equiv -\frac{\partial^2}{\partial x^2} + g(x)\eta^2$ and the eigenvalue problem

$$\begin{aligned}
L_\eta v(x,\eta) &= \lambda\, v(x,\eta), \qquad x \in I_a \equiv (-a,a), \\
v(a,\eta) &= v(-a,\eta) = 0.
\end{aligned}$$

The least eigenvalue is given by the Rayleigh quotient formula

$$\begin{aligned}\lambda_0(a,\eta) &= \inf_{f(\neq 0)\in C_0^\infty(I_a)} \frac{\langle L_\eta f, f\rangle_{L^2}}{\|f\|_{L^2}^2} \\ &= \inf_{f(\neq 0)\in C_0^\infty(I_a)} \frac{\int_{-a}^{a} f'(x)^2\,dx + \int_{-a}^{a} g(x)\,\eta^2 f(x)^2\,dx}{\|f\|_{L^2}^2},\end{aligned}$$

from which it follows that

$$\lambda_0(a,\eta) \le \lambda_0(a_0,\eta) \text{ if } a \ge a_0. \tag{9.2}$$

The decay condition (9.1) above is equivalent to the existence of $\delta_0 > 0$ such that $g(x) \le e^{-\frac{\delta_0}{|x|}}$ for x small. So we may suppose $g(x) \le Ce^{-\frac{\delta_0}{|x|}}$ for $x \in I \equiv [-1,1]$ where $C \ge 1$, and then take $|\eta|$ sufficiently large that with

$$a(\eta) \equiv \frac{\delta_0}{\ln C + 2\ln|\eta|},$$

we have both $a(\eta) \le 1$ and

$$g(x)\,\eta^2 \le Ce^{-\frac{\delta_0}{|x|}}\eta^2 \le e^{\ln C - \frac{\delta_0}{a(\eta)} + 2\ln|\eta|} = 1, \qquad x \in I_{a(\eta)}.$$

Now let $\mu_0(a(\eta))$ denote the least eigenvalue for the problem

$$\begin{aligned}\left\{-\frac{\partial^2}{\partial x^2} + 1\right\} v(x,\eta) &= \mu\, v(x,\eta), \qquad x \in I_{a(\eta)} = (-a(\eta), a(\eta)), \\ v(a(\eta),\eta) &= v(-a(\eta),\eta) = 0,\end{aligned}$$

and note that

$$\begin{aligned}\mu_0(a(\eta)) &= \inf_{f(\neq 0)\in C_0^\infty(I_{a(\eta)})} \frac{\left\langle -\frac{\partial^2 f}{\partial x^2} + f, f\right\rangle_{L^2(I_{a(\eta)})}}{\|f\|_{L^2(I_{a(\eta)})}^2} \\ &= \inf_{f(\neq 0)\in C_0^\infty(I_{a(\eta)})} \frac{\int_{-a(\eta)}^{a(\eta)} f'(x)^2\,dx + \int_{-a(\eta)}^{a(\eta)} f(x)^2\,dx}{\|f\|_{L^2(I_{a(\eta)})}^2}.\end{aligned}$$

It follows that

$$\lambda_0(a(\eta),\eta) \le \mu_0(a(\eta)), \qquad \text{for } |\eta| \text{ sufficiently large.}$$

Now an easy classical calculation using exact solutions to $\left\{-\frac{\partial^2}{\partial x^2} + 1 - \mu\right\} v = 0$ shows that

$$\mu_0(a(\eta)) = C_1 \frac{1}{a(\eta)^2} + 1,$$

for some constant C_1 independent of η, and hence combining this with (9.2), we have

$$\begin{aligned}0 < \lambda_0(1,\eta) &\le \lambda_0(a(\eta),\eta) \le \mu_0(a(\eta)) \\ &= C_1\left(\frac{\ln C + 2\ln|\eta|}{\delta_0}\right)^2 + 1 \le C_2(\ln|\eta|)^2, \qquad \text{for } |\eta| \text{ sufficiently large.}\end{aligned} \tag{9.3}$$

Now let $v_0(x,\eta)$ be an eigenfunction on the interval $I = -I_1 = [-1,1]$ associated with $\lambda_0(1,\eta)$ and normalized so that

$$\|v_0(\cdot,\eta)\|_{L^2(I)} = 1. \tag{9.4}$$

Choose a sequence $\{a_n\}_{n=-\infty}^{\infty}$ satisfying

$$|a_n| \leq \frac{1}{1+|n|^{\alpha}}\rho_n, \tag{9.5}$$

for some $\alpha > 0$ where $\left\|\{\rho_n\}_{n\in\mathbb{Z}}\right\|_{\ell^2} = 1$. For $y \in I_\pi = [-\pi, \pi]$, identified with the unit circle $\mathbb{T}$ upon identifying $-\pi$ and π, we formally define

$$\begin{aligned} w(x,y) &\equiv \sum_{n\in\mathbb{Z}} e^{iyn} v_0(x,n)\, a_n; \\ w_N(x,y) &\equiv \sum_{n\in\mathbb{Z}} \lambda_0(1,n)^N e^{iyn} v_0(x,n)\, a_n; \\ u(x,y,t) &\equiv \sum_{N=0}^{\infty} \frac{t^{2N}}{(2N)!} w_N(x,y). \end{aligned}$$

We now claim that

$$w_N(x,y) = \left\{-\frac{\partial^2}{\partial x^2} - g(x)\frac{\partial^2}{\partial y^2}\right\}^N w(x,y).$$

Indeed, assuming this holds for N, and using

$$-\frac{\partial^2}{\partial x^2} v_0(x,n) = \left[\lambda_0(1,n) - g(x)\, n^2\right] v_0(x,n),$$

we obtain that

$$\begin{aligned} \left\{-\frac{\partial^2}{\partial x^2} - g(x)\frac{\partial^2}{\partial y^2}\right\}^{N+1} w(x,y) &= \left\{-\frac{\partial^2}{\partial x^2} - g(x)\frac{\partial^2}{\partial y^2}\right\} w_N(x,y) \\ &= -\sum_{n\in\mathbb{Z}} \lambda_0(1,n)^N e^{iyn} \frac{\partial^2}{\partial x^2} v_0(x,n)\, a_n \\ &\quad + \sum_{n\in\mathbb{Z}} \lambda_0(1,n)^N e^{iyn} g(x)\, n^2 v_0(x,n)\, a_n \\ &= \sum_{n\in\mathbb{Z}} \lambda_0(1,n)^N e^{iyn} \lambda_0(1,n)\, v_0(x,n)\, a_n \\ &\quad - \sum_{n\in\mathbb{Z}} \lambda_0(1,n)^N e^{iyn} g(x)\, n^2 v_0(x,n)\, a_n \\ &\quad + \sum_{n\in\mathbb{Z}} \lambda_0(1,n)^N e^{iyn} g(x)\, n^2 v_0(x,n)\, a_n \\ &= \sum_{n\in\mathbb{Z}} \lambda_0(1,n)^{N+1} e^{iyn} v_0(x,n)\, a_n \\ &= w_{N+1}(x,y). \end{aligned}$$

It follows that

$$\left\{-\frac{\partial^2}{\partial x^2} - g(x)\frac{\partial^2}{\partial y^2}\right\} w_N(x,y) = w_{N+1}(x,y),$$

and so formally we get

$$
\begin{aligned}
Lu(x,y,t) &= \left\{-\frac{\partial^2}{\partial x^2} - g(x)\frac{\partial^2}{\partial y^2} - \frac{\partial^2}{\partial t^2}\right\}\sum_{N=0}^{\infty}\frac{t^{2N}}{(2N)!}w_N(x,y) \\
&= \sum_{N=0}^{\infty}\frac{t^{2N}}{(2N)!}w_{N+1}(x,y) - \sum_{N=0}^{\infty}\frac{\partial^2}{\partial t^2}\frac{t^{2N}}{(2N)!}w_N(x,y) \\
&= \sum_{N=0}^{\infty}\frac{t^{2N}}{(2N)!}w_{N+1}(x,y) - \sum_{N=1}^{\infty}\frac{t^{2N-2}}{(2N-2)!}w_N(x,y) = 0.
\end{aligned}
$$

Now we show that $u(x,y,t)$ is well defined as an $L^2(I\times\mathbb{T})$-valued analytic function of t for t in some small neighbourhood of 0 provided $\{a_n\}_{n\in\mathbb{Z}}$ is in $\ell^2(\mathbb{Z})$ with suitable decay at ∞, namely (9.5). Here $I=[-1,1]$. Indeed, using Plancherel's formula in the y variable, and then Fubini's theorem, we have

$$
\begin{aligned}
\|w_N\|^2_{L^2(I\times\mathbb{T})} &= \int_{-1}^{1}\left\{\int_{-\pi}^{\pi}\left|\sum_{n\in\mathbb{Z}}\lambda_0(1,n)^N e^{iyn}v_0(x,n)\,a_n\right|^2 dy\right\}dx \\
&= \int_{-1}^{1}\left\{\sum_{n\in\mathbb{Z}}\left|\lambda_0(1,n)^N v_0(x,n)\,a_n\right|^2\right\}dx \\
&= \sum_{n\in\mathbb{Z}}\left\{\int_{-1}^{1}|v_0(x,n)|^2\,dx\right\}\left|\lambda_0(1,n)^N a_n\right|^2 \\
&= \sum_{n\in\mathbb{Z}}\left|\lambda_0(1,n)^N a_n\right|^2.
\end{aligned}
$$

Now from (9.3) we have the bound $\lambda_0(1,n)\le C_2(\ln n)^2$ for n sufficiently large, and hence from (9.5),

$$
|a_n| \le \frac{1}{1+|n|^{\alpha}}\rho_n,
$$

where $\left\|\{\rho_n\}_{n\in\mathbb{Z}}\right\|_{\ell^2}=1$, we obtain

$$
\begin{aligned}
\|w_N\|_{L^2(I\times\mathbb{T})} &\le C_3\sqrt{\sum_{n\in\mathbb{Z}}\left|(\ln n)^{2N}a_n\right|^2} \\
&\le C_3\sqrt{\sum_{n\in\mathbb{Z}}\left|(\ln n)^{2N}\left(\frac{1}{1+e^{\alpha\ln n}}\right)\right|^2|\rho_n|^2} \\
&\le C_4\sqrt{N}\alpha^{-2N}(2N)!\sqrt{\sum_{n\in\mathbb{Z}}|\rho_n|^2} = C_4\frac{1}{\sqrt{N}}\alpha^{-2N}(2N)!
\end{aligned}
$$

since the maximum value of $s^{2N}e^{-\alpha s}$ occurs at $s=\frac{2N}{\alpha}$, and then by Stirling's formula,

$$
\frac{s^{2N}}{1+e^{\alpha s}} \le s^{2N}e^{-\alpha s} \le \left(\frac{2N}{\alpha}\right)^{2N}e^{-\alpha\frac{2N}{\alpha}} = \left(\frac{2N}{e}\right)^{2N}\alpha^{-2N} \le \frac{1}{\sqrt{N}}\alpha^{-2N}(2N)!
$$

Thus we conclude that

$$\|u(x,y,t)\|_{L^2_{x,y}(I\times\mathbb{T})} \le \sum_{N-0}^{\infty} \frac{t^{2N}}{(2N)!}\|w_N\|_{L^2(I\times\mathbb{T})} \le C_4 \sum_{N=0}^{\infty} \frac{1}{\sqrt{N}}\left(\frac{t}{\alpha}\right)^{2N} < \infty$$

for $t\in(-\alpha,\alpha)$, and it follows that $u(x,y,t)$ is a well-defined $L^2(I\times\mathbb{T})$-valued analytic function of $t\in(-\alpha,\alpha)$ that satisfies the homogeneous equation $\mathcal{L}u(x,y,t)=0$ for $(x,y,t)\in I\times\mathbb{T}\times(-\alpha,\alpha)$.

Next, we show that $u\in W^{1,2}_A(I\times\mathbb{T}\times(-\varepsilon,\varepsilon))$ for some $\varepsilon>0$. We first compute that

$$\begin{aligned}
&\left\|\frac{\partial}{\partial x}w_N\right\|^2_{L^2(I\times\mathbb{T})} + \left\|\sqrt{g(x)}\frac{\partial}{\partial y}w_N\right\|^2_{L^2(I\times\mathbb{T})}\\
=\;& \left\langle\left\{-\frac{\partial^2}{\partial x^2} - g(x)\frac{\partial^2}{\partial y^2}\right\} w_N(x,y), w_N(x,y)\right\rangle_{L^2(I\times\mathbb{T})}\\
=\;& \langle w_{N+1}(x,y), w_N(x,y)\rangle_{L^2(I\times\mathbb{T})}\\
\le\;& \|w_{N+1}\|_{L^2(I\times\mathbb{T})}\|w_N\|_{L^2(I\times\mathbb{T})}\\
\le\;& C_4\frac{1}{\sqrt{N+1}}\alpha^{-2N-2}(2N+2)!C_4\frac{1}{\sqrt{N+1}}\alpha^{-2N}(2N)!\\
\le\;& C_5^2 N\left[(2N)!\right]^2\alpha^{-4N-2},
\end{aligned}$$

which shows in particular that $w_N\in W^{1,2}_A(I\times\mathbb{T})$ for each $N\ge1$ with the norm estimate

$$\left\|\frac{w_N}{(2N)!}\right\|_{W^{1,2}_A(I\times\mathbb{T})} \le C_5\sqrt{N}\alpha^{-2N-1}.$$

Thus the $W^{1,2}_A(I\times\mathbb{T})$-valued analytic function $u(t)=\sum_{N=0}^{\infty}\frac{w_N}{(2N)!}t^{2N}$ is $W^{1,2}_A(I\times\mathbb{T})$-bounded in the complex disk $B(0,\alpha)$ centered at the origin with radius α. Then we use Cauchy's estimates for the $W^{1,2}_A(I\times\mathbb{T})$-valued analytic function $u(t)$ to obtain that $\frac{\partial}{\partial t}u(t)$ is $W^{1,2}_A(I\times\mathbb{T})$-bounded in any complex disk $B(0,\varepsilon)$ with $0<\varepsilon<\alpha=\beta-\frac{1}{2}$, which shows that $\frac{\partial}{\partial t}u\in L^2(I\times\mathbb{T}\times(-\varepsilon,\varepsilon))$ for $0<\varepsilon<\beta-\frac{1}{2}$. This completes the proof that $u\in W^{1,2}_A(I\times\mathbb{T}\times(-\varepsilon,\varepsilon))$ for some $\varepsilon>0$.

Finally, we note that with $\rho_n=\frac{1}{1+|n|^\beta}$ where $\frac{1}{2}<\beta\le\frac{3}{2}-\alpha$, then u is *not* smooth near the origin since

$$\begin{aligned}
\left\|\frac{\partial}{\partial y}w\right\|^2_{L^2(I\times\mathbb{T})} &= \int_{-1}^{1}\left\{\int_{-\pi}^{\pi}\left|\sum_{n\in\mathbb{Z}} ine^{iyn}v_0(x,n)a_n\right|^2 dy\right\}dx\\
&= \int_{-1}^{1}\left\{\sum_{n\in\mathbb{Z}}\left|v_0(x,n)na_n\right|^2\right\}dx\\
&= \sum_{n\in\mathbb{Z}}\left\{\int_{-1}^{1}|v_0(x,n)|^2dx\right\}|na_n|^2\\
&= \sum_{n\in\mathbb{Z}}|na_n|^2 = \sum_{n\in\mathbb{Z}}\left|\frac{n}{1+|n|^\alpha}\rho_n\right|^2 = \infty.
\end{aligned}$$

This is essentially the example of Morimoto [**Mor**]. However, we need more - namely, we must construct an essentially unbounded weak solution u in some neighbourhood $(-\delta,\delta)\times\mathbb{T}\times(-\varepsilon,\varepsilon)$.

To accomplish this, we first derive the additional property (9.6) below of the least eigenfunction $v_n(x)\equiv v_0(x,n)$ that satisfies the equation

$$\begin{aligned}\left\{-\frac{\partial^2}{\partial x^2}+g(x)\,n^2\right\}v_n(x) &= \lambda_0(1,n)\ v_n(x),\\ v_n(-1) &= v_n(1)=0.\end{aligned}$$

We claim that $v_n(x)$ is even on $[-1,1]$ and decreasing from $v_n(0)$ to 0 on the interval $[0,1]$. Indeed, the least eigenfunction v_n minimizes the Rayleigh quotient

$$\begin{aligned}&\frac{\int_{-1}^1 v_n'(x)^2\,dx+\int_{-1}^1 g(x)\,n^2 v_n(x)^2\,dx}{\|v_n\|_{L^2}^2}\\ &=\inf_{f(\neq 0)\in C_0^\infty(I_a)}\frac{\int_{-1}^1 f'(x)^2\,dx+\int_{-1}^1 g(x)\,n^2 f(x)^2\,dx}{\|f\|_{L^2}^2},\end{aligned}$$

and since the radially decreasing rearrangement v_n^* of v_n on $[-1,1]$ satisfies both

$$\int_{-1}^1 v_n^{*\prime}(x)^2\,dx\leq\int_{-1}^1 v_n'(x)^2\,dx$$

and

$$\int_{-1}^1 g(x)\,n^2 v_n^*(x)^2\,dx\leq\int_{-1}^1 g(x)\,n^2 v_n(x)^2\,dx,$$

as well as $\|v_n^*\|_{L^2}^2=\|v_n\|_{L^2}^2$, we conclude that $v_n=v_n^*$. The only simple consequence we need from this is that

$$2v_n(0)^2\geq\int_{-1}^1 v_n(x)^2\,dx=1,\qquad n\geq 1,\tag{9.6}$$

where the equality follows from our normalizing assumption $\|v_n\|_{L^2}=1$ in (9.4).

Now recall $\alpha>0$ from (9.5) above, and choose $0<\alpha<\alpha'\leq\frac{1}{4}$ and define

$$a_n=\left\{\begin{array}{ccc}\frac{1}{n^{\frac{1}{2}+\alpha'}} & \text{for} & n\geq 1\\ 0 & \text{for} & n\leq 0\end{array}\right..\tag{9.7}$$

Then for each $(x,t)\in I\times(-\alpha,\alpha)$, we have with

$$B_n(x,t)\equiv\left[\sum_{N=0}^\infty\frac{t^{2N}}{(2N)!}\lambda_0(1,n)^N\right]v_n(x)\,a_n,$$

that

$$\begin{aligned}u(x,y,t) &= \sum_{N=0}^\infty\frac{t^{2N}}{(2N)!}\sum_{n=1}^\infty\lambda_0(1,n)^N e^{iyn}v_n(x)\,a_n=\sum_{n=1}^\infty e^{iyn}B_n(x,t)\ ,\\ w(x,y)^2 &= \left(\sum_{n=1}^\infty e^{iyn}B_n(x,t)\right)^2=\sum_{n=2}^\infty\left\{\sum_{k=1}^{n-1}B_{n-k}(x,t)\,B_k(x,t)\right\}e^{iyn}\ ,\end{aligned}$$

and so by Plancherel's theorem,

$$\|u(x,\cdot,t)\|_{L^4(\mathbb{T})}^4 = \left\|u(x,\cdot,t)^2\right\|_{L^2(\mathbb{T})}^2 = \sum_{n=2}^{\infty}\left|\sum_{k=1}^{n-1} B_{n-k}(x,t)\,B_k(x,t)\right|^2 .$$

In particular we have from (9.6) that

$$\begin{aligned}
\|u(0,\cdot,0)\|_{L^4(\mathbb{T})}^4 &= \sum_{n=2}^{\infty}\left|\sum_{k=1}^{n-1} B_{n-k}(0,0)\,B_k(0,0)\right|^2 \\
&= \sum_{n=2}^{\infty}\left|\sum_{k=1}^{n-1} v_{n-k}(0)\,a_{n-k}v_k(0)\,a_k\right|^2 \geq \frac{1}{2}\sum_{n=2}^{\infty}\left|\sum_{k=1}^{n-1} a_{n-k}a_k\right|^2 ,
\end{aligned}$$

and now we obtain that $\|w(0,\cdot)\|_{L^4(\mathbb{T})}^4 = \infty$ from the estimates

$$\begin{aligned}
\sum_{k=1}^{n-1} a_{n-k}a_k &= \sum_{k=1}^{n-1}\frac{1}{(n-k)^{\frac{1}{2}+\alpha'}}\frac{1}{k^{\frac{1}{2}+\alpha'}} \gtrsim \frac{1}{\left(n-\frac{n}{2}\right)^{\frac{1}{2}+\alpha'}}\sum_{k=1}^{\frac{n}{2}}\frac{1}{k^{\frac{1}{2}+\alpha'}} \\
&\gtrsim \frac{1}{\left(\frac{n}{2}\right)^{\frac{1}{2}+\alpha'}}\left(\frac{n}{2}\right)^{\frac{1}{2}-\alpha'} \approx \frac{1}{n^{2\alpha'}}
\end{aligned}$$

and

$$\|w(0,\cdot)\|_{L^4(\mathbb{T})}^4 \geq \frac{1}{2}\sum_{n=2}^{\infty}\left|\sum_{k=1}^{n-1} a_{n-k}a_k\right|^2 \gtrsim \sum_{n=2}^{\infty}\frac{1}{n^{4\alpha'}} = \infty, \qquad \text{for } \alpha' \leq \frac{1}{4}.$$

Now we note that each eigenfunction $v_n(x)$ is continuous in x since it solves an elliptic second order equation on the interval $[-1,1]$. Moreover it now follows that $B_n(x,t)$ is jointly continuous on the rectangle $(-1,1)\times(-\alpha,\alpha)$. Then we write

$$\begin{aligned}
&\sum_{n=2}^{\infty}\left|\sum_{k=1}^{n-1} B_{n-k}(x,t)\,B_k(x,t)\right|^2 \\
&\qquad = \sum_{\substack{\beta=(\beta_1,\beta_2,\beta_3,\beta_4)\in\mathbb{N}^4\\ \beta_1+\beta_2=n=\beta_3+\beta_4}} B_{\beta_1}(x,t)\,B_{\beta_2}(x,t)\,B_{\beta_3}(x,t)\,B_{\beta_4}(x,t),
\end{aligned}$$

and apply Fatou's lemma to conclude that

$$\begin{aligned}
\infty &= \sum_{n=2}^{\infty}\left|\sum_{k=1}^{n-1} B_{n-k}(0,0)\, B_k(0,0)\right|^2 \\
&= \sum_{\substack{\beta=(\beta_1,\beta_2,\beta_3,\beta_4)\in\mathbb{N}^4\\ \beta_1+\beta_2=n=\beta_3+\beta_4}} B_{\beta_1}(0,0)\, B_{\beta_2}(0,0)\, B_{\beta_3}(0,0)\, B_{\beta_4}(0,0) \\
&= \sum_{\substack{\beta=(\beta_1,\beta_2,\beta_3,\beta_4)\in\mathbb{N}^4\\ \beta_1+\beta_2=n=\beta_3+\beta_4}} \liminf_{(x,t)\to(0,0)} \left\{B_{\beta_1}(x,t)\, B_{\beta_2}(x,t)\, B_{\beta_3}(x,t)\, B_{\beta_4}(x,t)\right\} \\
&\leq \liminf_{(x,t)\to(0,0)} \sum_{\substack{\beta=(\beta_1,\beta_2,\beta_3,\beta_4)\in\mathbb{N}^4\\ \beta_1+\beta_2=n=\beta_3+\beta_4}} B_{\beta_1}(x,t)\, B_{\beta_2}(x,t)\, B_{\beta_3}(x,t)\, B_{\beta_4}(x,t) \\
&= \liminf_{(x,t)\to(0,0)} \sum_{n=2}^{\infty}\left|\sum_{k=1}^{n-1} B_{n-k}(x,t)\, B_k(x,t)\right|^2 = \liminf_{(x,t)\to(0,0)} \|u(x,\cdot,t)\|^4_{L^4(\mathbb{T})}.
\end{aligned}$$

Thus we have $\lim_{(x,t)\to(0,0)} \|u(x,\cdot,t)\|^4_{L^4(\mathbb{T})} = \infty$, which implies that $\|u\|_{L^\infty((-\delta,\delta)\times\mathbb{T}\times(-\varepsilon',\varepsilon'))} = \infty$ for all $0<\delta<1$ and $0<\varepsilon'<\varepsilon<\alpha$. Thus we have shown that

$$u(x,y,t) = \sum_{N=0}^{\infty} \frac{t^{2N}}{(2N)!} w_N(x,y) = w(x,y) + \sum_{N=1}^{\infty} \frac{t^{2N}}{(2N)!} w_N(x,y)$$

is a weak solution of $Lu=0$ in $I\times\mathbb{T}\times(-\varepsilon,\varepsilon)$ with $\|u\|_{L^\infty((-\delta,\delta)\times\mathbb{T}\times(-\varepsilon',\varepsilon'))} = \infty$ provided $0<\delta<1$ and $0<\varepsilon'<\varepsilon<\alpha<\alpha'\leq\frac{1}{4}$. This completes the proof that u fails to be essentially bounded on $(-\delta,\delta)\times\mathbb{T}\times(-\varepsilon',\varepsilon')$. □

Part 4

Appendix

Our first result here is the quasilinear hypoellipticity theorem in higher dimensions mentioned in the introduction. Then we include two results tangential to our development here. We show that our hypoellipticity theorem for quasilinear equations in the plane doesn't generalize to more fully nonlinear equations, even with a degeneracy like that in (2.4) above. Then we compute the Fedii operator $\mathcal{L}$ in metric polar coordinates, and show that there are no nonconstant radial functions u for which $\mathcal{L}u$ is also radial.

APPENDIX A

A Quasilinear Hypoellipticity Theorem

Let $n \geq 2$ and let $\Omega \subset \mathbb{R}^n$ be a bounded domain, $\Gamma = \Omega \times \mathbb{R}$ and let

$$\Lambda(x) = \begin{bmatrix} 1 & 0 & \dots & & 0 \\ 0 & 1 & & & 0 \\ \vdots & & \ddots & & \vdots \\ & & & 1 & 0 \\ 0 & 0 & \dots & 0 & \tilde{f}(x') \end{bmatrix}$$

where $x' = x_1, \dots, x_{n-1}$ and $f(x) = f(x')$ is in $C^2(\Omega)$. Let $\mathcal{A}(x,z)$ be a smooth $n \times n$ symmetric matrix function, and let $g(x,z)$ be a smooth function in Γ. We will assume the following:

(1) f is nonnegative, $f \leq 1$, and $f(x) = f(x') > 0$ if $x' \neq 0$;

(2) $f(r) = \inf\left\{\tilde{f}(x) : |x_1| = r \geq 0,\ x = (x_1, x_2, \dots, x_n) \in \Omega\right\}$ satisfies $f(r) \gtrsim e^{-F(r)}$, where $F(r) = F_{3,\sigma}(r) = \left(\ln \frac{1}{r}\right)\left(\ln\ln\ln \frac{1}{r}\right)^{\sigma}$ for some $0 < \sigma < 1$.

(3) (Diagonal condition in [**RSaW2**]) For some $C > 0$

$$C^{-1}\xi \cdot \Lambda(x)\xi \leq \xi \cdot \mathcal{A}(x,z)\xi \leq C\,\xi \cdot \Lambda(x)\xi \tag{A.1}$$

(4) (super-subordination condition in [**RSaW2**]) For some $B > 0$

$$\frac{|\partial_z \mathcal{A}(x,z)\xi|^2}{f(x)\,\xi^t \mathcal{A}(x,z)\xi} = o(f(x)) \qquad \text{as } x \to 0, \tag{A.2}$$

$$\sum_{i=1}^{n} |\partial_i \partial_z \mathcal{A}(x,z)\xi|^2 + \left|\partial_z^2 \mathcal{A}(x,z)\xi\right|^2 \leq B^2\,\xi^t \mathcal{A}(x,z)\xi. \tag{A.3}$$

(5) $g(x,z)\operatorname{sign} z \leq 0$ and $g_z(x,z) \leq 0$ in Γ.

The proof of the next theorem is a routine application of our geometric continuity Theorem 84 with the main result in [**RSaW1**]. See also [**RSaW2**] for applications to the Dirichlet problem for the Monge-Ampère equation.

THEOREM 86 (Regularity of solutions). *Let $\Omega \subset \mathbb{R}^n$ be a bounded domain and $\Gamma = \Omega \times \mathbb{R}$. Suppose that $f \in \mathcal{C}^2(\Omega)$ satisfies (1) and (2), that $\mathcal{A}(x,z)$, $g(x,z) \in \mathcal{C}^\infty(\Gamma)$, $\mathcal{A}$ satisfies the diagonal condition (3) and the super subordination Condition (4) in Γ, and g satisfies (5) in Γ. Then, any weak solution w of*

$$\operatorname{div} \mathcal{A}(x,w)\nabla w + g(x,w) = 0 \qquad \text{in } \Omega$$

is also a strong solution and satisfies $w \in \mathcal{C}^\infty(\Omega)$.

APPENDIX B

A Monge-Ampère Example

Let $\varphi(s)$ be a smooth even strictly convex function that vanishes only at $s = 0$, and vanishes to infinite order there. Then we have

$$|\varphi'(s)|^2 \leq \|\varphi''\|_\infty \varphi(s) \ll \varphi(s).$$

Now define $u(x,y) \equiv x^2 + \varphi(x)y^2$ and compute

$$\begin{aligned} D^2u(x,y) &= \begin{bmatrix} u_{xx} & u_{xy} \\ u_{yx} & u_{yy} \end{bmatrix} = \begin{bmatrix} 2+\varphi''(x)y^2 & 2\varphi'(x)y \\ 2\varphi'(x)y & 2\varphi(x) \end{bmatrix}; \\ \det D^2u(x,y) &= 4\varphi(x) + 2\varphi(x)\varphi''(x)y^2 - 4\varphi'(x)^2y^2. \end{aligned}$$

Then with

$$f(x,y) \equiv \left(2+\varphi''(x)y^2\right)2\varphi(x) - 4\varphi'(x)^2y^2$$

we have for (x,y) small enough that u is a convex solution to the Monge-Ampère equation

$$\det D^2u(x,y) = f(x,y) \tag{B.1}$$

where f is smooth and positive away from $y = 0$ and satisfies

$$\frac{\partial}{\partial y} f(x,y) = o(f(x,y)).$$

Now we modify u by changing the multiple of x^2 on either side of the y-axis, which has little effect on $\det D^2u(x,y)$:

$$\begin{aligned} u(x,y) &\equiv \begin{cases} x^2+\varphi(x)y^2 & \text{if} \quad x \geq 0 \\ \frac{1}{2}x^2+\varphi(x)y^2 & \text{if} \quad x \leq 0 \end{cases}, \\ \det D^2u(x,y) &= \begin{cases} \left(2+\varphi''(x)y^2\right)2\varphi(x) - 4\varphi'(x)^2y^2 & \text{if} \quad x \geq 0 \\ \left(1+\varphi''(x)y^2\right)2\varphi(x) - \varphi'(x)^2y^2 & \text{if} \quad x \leq 0 \end{cases}. \end{aligned}$$

Thus with $f(x,y) \equiv \det D^2u(x,y)$, we still have that f is smooth and positive away from $y = 0$ and satisfies

$$\frac{\partial}{\partial y} f(x,y) = o(f(x,y)).$$

But now $u \in C^{1,1} \setminus C^2$ is a nonsmooth solution to the Monge-Ampère equation $\det D^2u = f$. Of course u is constant on the y axis, and the existence of this 'Pogorelov segment' accounts for the singularity of the solution - see [**SaW**].

The partial Legendre transform exhibits a close connection between this equation and quasilinear equations of the type considered in Theorem 1. Indeed, if u solves (B.1), then the associated partial Legendre transform

$$\begin{aligned} s &= x \text{ and } t = u_y(x,y), \\ z &= u_x(x,y) \text{ and } v = y, \end{aligned}$$

satisfies the 'Cauchy-Riemann' equations,

$$\begin{aligned} z_s &= f v_t , \\ z_t &= -v_s , \end{aligned}$$

and hence v is a weak solution of the quasilinear equation

$$\left\{ \frac{\partial^2}{\partial s^2} + \frac{\partial}{\partial t} f\left(s, v\left(s,t\right)\right) \frac{\partial}{\partial t} \right\} v = 0,$$

of the form $\mathcal{L}_{\text{quasi}}$ in Theorem 1. The transform has nonnegative Jacobian

$$\frac{\partial\left(s,t\right)}{\partial\left(x,y\right)} = \det \begin{bmatrix} 1 & 0 \\ u_{yx} & u_{yy} \end{bmatrix} = u_{yy} ,$$

which is positive where f is positive. But $kv_t = z_s = u_{xx}$ has a discontinuity on the t-axis, and it follows easily that both $v_t = \frac{1}{f} z_s$ blows up at the t-axis, and that v has a discontinuity across the t-axis. Of course $v = y$ is bounded. The resolution here is that the partial Legendre transform is not one-to-one on the y-axis, and in fact the transformed equation is not valid at $x = 0$.

APPENDIX C

Absence of Radial Solutions

Recall our family of geometries with inverse metric tensor $A = \begin{bmatrix} 1 & 0 \\ 0 & f(x)^2 \end{bmatrix}$ and A-distance dt given by $dt^2 = dx^2 + \frac{1}{f(x)^2} dy^2$, which coincides with the familiar control metric d associated with A. Here we suppose that $f(x)$ is positive away from zero and $f(0) = 0$ (thus prohibiting the usual elliptic geometry). We say that a function $v(x, y)$ is radial if it depends only on the metric distance $r = d((0,0),(x,y))$ from the point (x, y) to the origin. Here we show that there is **no** nonconstant radial solution to the equation $\mathcal{L}v = \varphi(r)$ with radial right hand side for such geometries. Note that this includes all of the finite type geometries of this form, as well as the infinitely degenerate ones.

For convenience we work in Region 1 as defined in Section 3 of Chapter 1 of Part 3. Recall that

$$\begin{aligned} \frac{\partial(x,y)}{\partial(r,\lambda)} &= \begin{bmatrix} \frac{\sqrt{\lambda^2 - f(x)^2}}{\lambda} & \frac{\sqrt{\lambda^2 - f(x)^2}}{\lambda} \cdot \int_0^x \frac{f(u)^2}{\left(\lambda^2 - f(u)^2\right)^{\frac{3}{2}}} du \\ \frac{f(x)^2}{\lambda} & \frac{f(x)^2 - \lambda^2}{\lambda} \cdot \int_0^x \frac{f(u)^2}{\left(\lambda^2 - f(u)^2\right)^{\frac{3}{2}}} du \end{bmatrix} \\ &= \begin{bmatrix} \frac{\sqrt{\lambda^2 - f(x)^2}}{\lambda} & \frac{\sqrt{\lambda^2 - f(x)^2}}{\lambda} m_3(x) \\ \frac{f(x)^2}{\lambda} & \frac{f(x)^2 - \lambda^2}{\lambda} m_3(x) \end{bmatrix}, \end{aligned}$$

where we write

$$m_k(x) = \int_0^x \frac{f(u)^2}{\left(\lambda^2 - f(u)^2\right)^{\frac{k}{2}}} du. \tag{C.1}$$

Then $\det\left(\frac{\partial(x,y)}{\partial(r,\lambda)}\right) = -\sqrt{\lambda^2 - f(x)^2} m_3(x)$, and the inverse matrix is given by

$$\begin{aligned} \frac{\partial(r,\lambda)}{\partial(x,y)} &= \frac{1}{\det \frac{\partial(x,y)}{\partial(r,\lambda)}} \begin{bmatrix} \frac{\sqrt{\lambda^2 - f(x)^2}}{\lambda} & \frac{\sqrt{\lambda^2 - f(x)^2}}{\lambda} m_3(x) \\ \frac{f(x)^2}{\lambda} & \frac{f(x)^2 - \lambda^2}{\lambda} m_3(x) \end{bmatrix} \\ &= \begin{bmatrix} \frac{\sqrt{\lambda^2 - f(x)^2}}{\lambda} & \frac{1}{\lambda} \\ \frac{f(x)^2}{\lambda\sqrt{\lambda^2 - f(x)^2} m_3(x)} & -\frac{1}{\lambda m_3(x)} \end{bmatrix}. \end{aligned}$$

Thus

$$\frac{\partial r}{\partial x} = \frac{\sqrt{\lambda^2 - f(x)^2}}{\lambda} \qquad \frac{\partial r}{\partial y} = \frac{1}{\lambda}$$

$$\frac{\partial \lambda}{\partial x} = \frac{f(x)^2}{\sqrt{\lambda^2 - f(x)^2}\lambda m_3(x)} \qquad \frac{\partial \lambda}{\partial y} = -\frac{1}{\lambda m_3(x)},$$

and so

$$\begin{aligned}
\frac{\partial}{\partial x} &= \frac{\partial \lambda}{\partial x}\frac{\partial}{\partial \lambda} + \frac{\partial r}{\partial x}\frac{\partial}{\partial r} \\
&= \frac{f(x)^2}{\sqrt{\lambda^2 - f(x)^2}\lambda m_3(x)}\frac{\partial}{\partial \lambda} + \frac{\sqrt{\lambda^2 - f(x)^2}}{\lambda}\frac{\partial}{\partial r}, \quad \text{and} \\
\frac{\partial}{\partial y} &= \frac{\partial \lambda}{\partial y}\frac{\partial}{\partial \lambda} + \frac{\partial r}{\partial y}\frac{\partial}{\partial r} = -\frac{1}{\lambda \int_0^x \frac{f(u)^2}{\left(\lambda^2 - f(u)^2\right)^{\frac{3}{2}}} du}\frac{\partial}{\partial \lambda} + \frac{1}{\lambda}\frac{\partial}{\partial r}.
\end{aligned}$$

REMARK 87. If $v = v(r)$ is radial, then

$$\begin{aligned}
\nabla_A v &= \left(\frac{\partial v}{\partial x}, f(x)\frac{\partial v}{\partial y}\right) = \left(\frac{\sqrt{\lambda^2 - f(x)^2}}{\lambda}\frac{\partial v}{\partial r}, f(x)\frac{1}{\lambda}\frac{\partial v}{\partial r}\right) \\
&= \left(\sqrt{1 - \left(\frac{f(x)}{\lambda}\right)^2}, \frac{f(x)}{\lambda}\right)\frac{\partial v}{\partial r},
\end{aligned}$$

which is not in general radial, but its modulus is:

$$|\nabla_A v| = \left|\frac{\partial v}{\partial r}\right|.$$

Note also that $\frac{f(x)}{\lambda} = \frac{f(x)}{f(X(\lambda))}$ where $(X(\lambda), Y(\lambda))$ is the turning point of the geodesic through the origin and (x, y).

Now we compute the second derivatives:

$$\begin{aligned}
\frac{\partial^2 r}{\partial x^2} &= \frac{\partial}{\partial x}\frac{\sqrt{\lambda^2 - f(x)^2}}{\lambda} \\
&= -\frac{\sqrt{\lambda^2 - f(x)^2}}{\lambda^2}\frac{\partial \lambda}{\partial x} + \frac{\lambda\frac{\partial \lambda}{\partial x} - f(x) f'(x)}{\lambda\sqrt{\lambda^2 - f(x)^2}} \\
&= -\frac{f(x)^2}{\lambda^3 m_3(x)} + \frac{f(x)^2}{\lambda\left(\lambda^2 - f(x)^2\right) m_3(x)} - \frac{f(x) f'(x)}{\lambda\sqrt{\lambda^2 - f(x)^2}}
\end{aligned}$$

and

$$\frac{\partial^2 r}{\partial y^2} = \frac{\partial}{\partial y}\frac{1}{\lambda} = -\frac{1}{\lambda^2}\frac{\partial \lambda}{\partial y} = \frac{1}{\lambda^3 m_3(x)}.$$

We then have

$$
\begin{aligned}
\frac{\partial^2 v}{\partial x^2} &= \frac{\partial}{\partial x}\left(\frac{\partial \lambda}{\partial x} v_\lambda (r,\lambda) + \frac{\partial r}{\partial x} v_r (r,\lambda)\right) \\
&= \frac{\partial \lambda}{\partial x}\frac{\partial}{\partial x} v_\lambda (r,\lambda) + \frac{\partial r}{\partial x}\frac{\partial}{\partial x} v_r (r,\lambda) \\
&\quad + \frac{\partial^2 \lambda}{\partial x^2} v_\lambda (r,\lambda) + \frac{\partial^2 r}{\partial x^2} v_r (r,\lambda) \\
&= \frac{\partial \lambda}{\partial x}\left(\frac{\partial \lambda}{\partial x}\frac{\partial}{\partial \lambda} + \frac{\partial r}{\partial x}\frac{\partial}{\partial r}\right) v_\lambda (r,\lambda) + \frac{\partial r}{\partial x}\left(\frac{\partial \lambda}{\partial x}\frac{\partial}{\partial \lambda} + \frac{\partial r}{\partial x}\frac{\partial}{\partial r}\right) v_r (r,\lambda) \\
&\quad + \frac{\partial^2 \lambda}{\partial x^2} v_\lambda (r,\lambda) + \frac{\partial^2 r}{\partial x^2} v_r (r,\lambda) \\
&= \left(\frac{\partial \lambda}{\partial x}\right)^2 v_{\lambda\lambda} (r,\lambda) + 2\frac{\partial r}{\partial x}\frac{\partial \lambda}{\partial x} v_{r\lambda} (r,\lambda) + \left(\frac{\partial r}{\partial x}\right)^2 v_{rr} (r,\lambda) \\
&\quad + \frac{\partial^2 \lambda}{\partial x^2} v_\lambda (r,\lambda) + \frac{\partial^2 r}{\partial x^2} v_r (r,\lambda),
\end{aligned}
$$

and similarly,

$$
\begin{aligned}
\frac{\partial^2 v}{\partial y^2} &= \left(\frac{\partial \lambda}{\partial y}\right)^2 v_{\lambda\lambda} (r,\lambda) + 2\frac{\partial r}{\partial y}\frac{\partial \lambda}{\partial y} v_{r\lambda} (r,\lambda) + \left(\frac{\partial r}{\partial y}\right)^2 v_{rr} (r,\lambda) \\
&\quad + \frac{\partial^2 \lambda}{\partial y^2} v_\lambda (r,\lambda) + \frac{\partial^2 r}{\partial y^2} v_r (r,\lambda).
\end{aligned}
$$

Then, if $v = v(r,\lambda)$, we have the following expression for $\mathcal{L}$ in polar coordinates:

$$
\begin{aligned}
\mathcal{L}v &= \frac{\partial^2 v}{\partial x^2} + f(x)^2 \frac{\partial^2 v}{\partial y^2} \\
&= \left(\left(\frac{\partial \lambda}{\partial x}\right)^2 + f(x)^2 \left(\frac{\partial \lambda}{\partial y}\right)^2\right) v_{\lambda\lambda} (r,\lambda) \\
&\quad + \left(\left(\frac{\partial r}{\partial x}\right)^2 + f(x)^2 \left(\frac{\partial r}{\partial y}\right)^2\right) v_{rr} (r,\lambda) \\
&\quad + 2\left(\frac{\partial r}{\partial x}\frac{\partial \lambda}{\partial x} + f(x)^2 \frac{\partial r}{\partial y}\frac{\partial \lambda}{\partial y}\right) v_{r\lambda} (r,\lambda) \\
&\quad + (\mathcal{L}\lambda)\ v_\lambda (r,\lambda) + (\mathcal{L}r)\ v_r (r,\lambda).
\end{aligned}
$$

Since

$$\left(\frac{\partial \lambda}{\partial x}\right)^2 + f(x)^2\left(\frac{\partial \lambda}{\partial y}\right)^2 = \left(\frac{f(x)^2}{\sqrt{\lambda^2 - f(x)^2}\lambda m_3(x)}\right)^2 + f(x)^2\left(-\frac{1}{\lambda m_3(x)}\right)^2$$
$$= \frac{f(x)^4}{\left(\lambda^2 - f(x)^2\right)\lambda^2 m_3(x)^2} + \frac{f(x)^2}{\lambda^2 m_3(x)^2}$$
$$= \frac{f(x)^4 + \left(\lambda^2 - f(x)^2\right) f(x)^2}{\left(\lambda^2 - f(x)^2\right)\lambda^2 m_3(x)^2}$$
$$= \frac{f(x)^2}{\left(\lambda^2 - f(x)^2\right) m_3(x)^2},$$

$$\left(\frac{\partial r}{\partial x}\right)^2 + f(x)^2\left(\frac{\partial r}{\partial y}\right)^2 = \left(\frac{\sqrt{\lambda^2 - f(x)^2}}{\lambda}\right)^2 + f(x)^2\left(\frac{1}{\lambda}\right)^2$$
$$= \frac{\lambda^2 - f(x)^2}{\lambda^2} + \frac{f(x)^2}{\lambda^2} = 1,$$

$$\frac{\partial r}{\partial x}\frac{\partial \lambda}{\partial x} + f(x)^2\frac{\partial r}{\partial y}\frac{\partial \lambda}{\partial y} = \frac{\sqrt{\lambda^2 - f(x)^2}}{\lambda}\frac{f(x)^2}{\sqrt{\lambda^2 - f(x)^2}\lambda m_3(x)} - f(x)^2\frac{1}{\lambda}\frac{1}{\lambda m_3(x)}$$
$$= \frac{f(x)^2}{\lambda^2 m_3(x)} - \frac{f(x)^2}{\lambda^2 m_3(x)} = 0,$$

we obtain

$$\begin{aligned}\mathcal{L}v &= \frac{\partial^2 v}{\partial x^2} + f(x)^2\frac{\partial^2 v}{\partial y^2} \\ &= \frac{f(x)^2}{\left(\lambda^2 - f(x)^2\right) m_3(x)^2} v_{\lambda\lambda}(r,\lambda) + v_{rr}(r,\lambda) \\ &\quad + (\mathcal{L}\lambda)\ v_\lambda(r,\lambda) + (\mathcal{L}r)\ v_r(r,\lambda).\end{aligned}$$

In particular, if $v = v(r)$ is radial,

$$\mathcal{L}v(r) = v_{rr}(r) + (\mathcal{L}r)\ v_r(r). \tag{C.2}$$

In order to understand equation (C.2), we must compute $\mathcal{L}r$,

$$\mathcal{L}r = \frac{\partial^2 r}{\partial x^2} + f(x)^2 \frac{\partial^2 r}{\partial y^2} = \left(\frac{\partial}{\partial x}\frac{\sqrt{\lambda^2 - f(x)^2}}{\lambda}\right) + \left(\frac{\partial}{\partial y}\frac{f(x)^2}{\lambda}\right) \tag{C.3}$$

$$= \frac{\lambda\frac{\partial \lambda}{\partial x} - f(x) f'(x)}{\lambda\sqrt{\lambda^2 - f(x)^2}} - \frac{\partial \lambda}{\partial x}\frac{\sqrt{\lambda^2 - f(x)^2}}{\lambda^2} - \frac{\partial \lambda}{\partial y}\frac{f(x)^2}{\lambda^2}$$

$$= \frac{\lambda \frac{f(x)^2}{\sqrt{\lambda^2 - f(x)^2}\lambda m_3(x)} - f(x) f'(x)}{\lambda\sqrt{\lambda^2 - f(x)^2}} - \frac{1}{\lambda}\left(\frac{\partial \lambda}{\partial x}\frac{\partial r}{\partial x} + f(x)^2 \frac{\partial \lambda}{\partial y}\frac{\partial r}{\partial y}\right)$$

$$= \frac{f(x)^2 - f(x) f'(x) m_3(x)\sqrt{\lambda^2 - f(x)^2}}{\lambda\left(\lambda^2 - f(x)^2\right) m_3(x)}$$

$$= \frac{f(x)^2}{\lambda\left(\lambda^2 - f(x)^2\right) m_3(x)} - \frac{f(x) f'(x)}{\lambda\sqrt{\lambda^2 - f(x)^2}},$$

where we used that $\frac{\partial \lambda}{\partial x}\frac{\partial r}{\partial x} + f(x)^2 \frac{\partial \lambda}{\partial y}\frac{\partial r}{\partial y} = 0$. Now we note that since

$$\frac{\partial}{\partial r} = \frac{\sqrt{\lambda^2 - f(x)^2}}{\lambda}\frac{\partial}{\partial x} + \frac{f(x)^2}{\lambda}\frac{\partial}{\partial y},$$

we have

$$\frac{\partial}{\partial r}\left(\sqrt{\lambda^2 - f(x)^2}\right)$$

$$= \frac{\sqrt{\lambda^2 - f(x)^2}}{\lambda}\frac{\partial}{\partial x}\left(\sqrt{\lambda^2 - f(x)^2}\right) + \frac{f(x)^2}{\lambda}\frac{\partial}{\partial y}\left(\sqrt{\lambda^2 - f(x)^2}\right)$$

$$= \frac{\sqrt{\lambda^2 - f(x)^2}}{\lambda}\frac{\lambda\frac{\partial \lambda}{\partial x} - f(x) f'(x)}{\sqrt{\lambda^2 - f(x)^2}} + \frac{f(x)^2}{\lambda}\frac{\lambda\frac{\partial \lambda}{\partial y}}{\sqrt{\lambda^2 - f(x)^2}}$$

$$= \frac{1}{\lambda}\left(\lambda\frac{f(x)^2}{\sqrt{\lambda^2 - f(x)^2}\lambda m_3(x)} - f(x) f'(x)\right) - \frac{f(x)^2}{\lambda}\frac{\lambda\frac{1}{\lambda m_3(x)}}{\sqrt{\lambda^2 - f(x)^2}}$$

$$= \frac{f(x)^2}{\sqrt{\lambda^2 - f(x)^2}\lambda m_3(x)} - \frac{f(x) f'(x)}{\lambda} - \frac{f(x)^2}{\lambda\sqrt{\lambda^2 - f(x)^2} m_3(x)}$$

$$= -\frac{f(x) f'(x)}{\lambda},$$

so that

$$\frac{\partial}{\partial r}\left(\ln\sqrt{\lambda^2 - f(x)^2}\right) = -\frac{f(x) f'(x)}{\lambda\sqrt{\lambda^2 - f(x)^2}}. \tag{C.4}$$

We also have

$$\begin{aligned}
\frac{\partial m_3(x)}{\partial x} &= \frac{\partial}{\partial x}\int_0^x \frac{f(u)^2}{\left(\lambda^2 - f(u)^2\right)^{\frac{3}{2}}}du \\
&= \frac{f(x)^2}{\left(\lambda^2 - f(x)^2\right)^{\frac{3}{2}}} - 3\lambda\frac{\partial\lambda}{\partial x}\int_0^x \frac{f(u)^2}{\left(\lambda^2 - f(u)^2\right)^{\frac{5}{2}}}du \\
&= \frac{f(x)^2}{\left(\lambda^2 - f(x)^2\right)^{\frac{3}{2}}} - 3\frac{f(x)^2 m_5(x)}{\sqrt{\lambda^2 - f(x)^2}m_3(x)}, \qquad \text{(C.5)}
\end{aligned}$$

$$\begin{aligned}
\frac{\partial m_3(x)}{\partial y} &= \frac{\partial}{\partial y}\int_0^x \frac{f(u)^2}{\left(\lambda^2 - f(u)^2\right)^{\frac{3}{2}}}du \\
&= -3\lambda\frac{\partial\lambda}{\partial y}\int_0^x \frac{f(u)^2}{\left(\lambda^2 - f(u)^2\right)^{\frac{5}{2}}}du \\
&= 3\frac{m_5(x)}{m_3(x)}, \qquad \text{(C.6)}
\end{aligned}$$

and hence

$$\begin{aligned}
\frac{\partial}{\partial r}m_3(x) &= \frac{\sqrt{\lambda^2 - f(x)^2}}{\lambda}\frac{\partial m_3(x)}{\partial x} + \frac{f(x)^2}{\lambda}\frac{\partial m_3(x)}{\partial y} \\
&= \frac{\sqrt{\lambda^2 - f(x)^2}}{\lambda}\frac{f(x)^2}{\left(\lambda^2 - f(x)^2\right)^{\frac{3}{2}}} - 3\frac{\sqrt{\lambda^2 - f(x)^2}}{\lambda}m_5(x)\lambda\frac{\partial\lambda}{\partial x} \\
&\quad -3\frac{f(x)^2}{\lambda}m_5(x)\lambda\frac{\partial\lambda}{\partial y} \\
&= \frac{\sqrt{\lambda^2 - f(x)^2}}{\lambda}\frac{f(x)^2}{\left(\lambda^2 - f(x)^2\right)^{\frac{3}{2}}} \\
&\quad -3\frac{\sqrt{\lambda^2 - f(x)^2}}{\lambda}m_5(x)\lambda\frac{f(x)^2}{\sqrt{\lambda^2 - f(x)^2}\lambda m_3(x)} \\
&\quad +3\frac{f(x)^2}{\lambda}m_5(x)\lambda\frac{1}{\lambda m_3(x)} \\
&= \frac{f(x)^2}{\lambda\left(\lambda^2 - f(x)^2\right)} - 3\frac{f(x)^2 m_5(x)}{\lambda m_3(x)} + 3\frac{f(x)^2 m_5(x)}{\lambda m_3(x)} \\
&= \frac{f(x)^2}{\lambda\left(\lambda^2 - f(x)^2\right)},
\end{aligned}$$

so that

$$\frac{\partial}{\partial r}\ln\left(m_3(x)\right) = \frac{f(x)^2}{\lambda\left(\lambda^2 - f(x)^2\right)m_3(x)}. \tag{C.7}$$

From (C.3), (C.4), and (C.7), we finally obtain

$$\begin{aligned}\mathcal{L}r &= \frac{\partial}{\partial r}\ln\left(m_3(x)\right) + \frac{\partial}{\partial r}\left(\ln\sqrt{\lambda^2 - f(x)^2}\right) \\ &= \frac{\partial}{\partial r}\ln\left(\sqrt{\lambda^2 - f(x)^2}m_3(x)\right).\end{aligned} \tag{C.8}$$

Substituting this into (C.2), we see that if $v(r)$ is a radial solution of $\mathcal{L}v = \varphi(r,\lambda)$, then

$$\frac{1}{\sqrt{\lambda^2 - f(x)^2}m_3(x)}\frac{\partial}{\partial r}\left(\sqrt{\lambda^2 - f(x)^2}m_3(x)\frac{\partial v}{\partial r}\right) = \varphi(r,\lambda). \tag{C.9}$$

CONCLUSION 88. If $v = v(r)$ is a radial solution of $\mathcal{L}v = 0$, then from (C.9) it follows that for some constant C,

$$\frac{\partial v}{\partial r} = \frac{C}{\sqrt{\lambda^2 - f(x)^2}m_3(x)},$$

and if $C \neq 0$, the function on the right is **not** radial. This means that there exist **no nonconstant radial solutions to** $\mathcal{L}v = 0$. Indeed, there is no nonconstant radial solution to $\mathcal{L}v = \varphi(r)$, since if v is a nonconstant radial solution, then from $\mathcal{L}v(r) = v''(r) + (\mathcal{L}r)\ v'(r)$ we see that $\mathcal{L}r$ is radial, and hence from (C.8) that $\sqrt{\lambda^2 - f(x)^2}m_3(x)$ is radial, a contradiction.

Bibliography

[BoGo] M. S. Baouendi and C. Goulaouic, *Nonanalytic-hypoellipticity for some degenerate elliptic operators*, Bull. Amer. Math. Soc. **78** (1972), 483–486, DOI 10.1090/S0002-9904-1972-12955-0. MR0296507

[CaVa] Luis A. Caffarelli and Alexis F. Vasseur, *The De Giorgi method for regularity of solutions of elliptic equations and its applications to fluid dynamics*, Discrete Contin. Dyn. Syst. Ser. S **3** (2010), no. 3, 409–427, DOI 10.3934/dcdss.2010.3.409. MR2660718

[Chr] Michael Christ, *Hypoellipticity in the infinitely degenerate regime*, Complex analysis and geometry (Columbus, OH, 1999), Ohio State Univ. Math. Res. Inst. Publ., vol. 9, de Gruyter, Berlin, 2001, pp. 59–84. MR1912731

[DeG] Ennio De Giorgi, *Sulla differenziabilità e l'analiticità delle estremali degli integrali multipli regolari* (Italian), Mem. Accad. Sci. Torino. Cl. Sci. Fis. Mat. Nat. (3) **3** (1957), 25–43. MR0093649

[Fe] V. S. Fediĭ, *On a criterion for hypoellipticity*, Math. USSR Sbornik **14** (1971), 15-45.

[FePh] C. Fefferman and D. H. Phong, *Subelliptic eigenvalue problems*, Conference on harmonic analysis in honor of Antoni Zygmund, Vol. I, II (Chicago, Ill., 1981), Wadsworth Math. Ser., Wadsworth, Belmont, CA, 1983, pp. 590–606. MR730094

[Fr] Bruno Franchi, *Weighted Sobolev-Poincaré inequalities and pointwise estimates for a class of degenerate elliptic equations*, Trans. Amer. Math. Soc. **327** (1991), no. 1, 125–158, DOI 10.2307/2001837. MR1040042

[FrLan1] Bruno Franchi and Ermanno Lanconelli, *De Giorgi's theorem for a class of strongly degenerate elliptic equations* (English, with Italian summary), Atti Accad. Naz. Lincei Rend. Cl. Sci. Fis. Mat. Natur. (8) **72** (1982), no. 5, 273–277 (1983). MR728257

[FrLan2] Bruno Franchi and Ermanno Lanconelli, *Hölder regularity theorem for a class of linear nonuniformly elliptic operators with measurable coefficients*, Ann. Scuola Norm. Sup. Pisa Cl. Sci. (4) **10** (1983), no. 4, 523–541. MR753153

[FrLan3] Bruno Franchi and Ermanno Lanconelli, *Une métrique associée à une classe d'opérateurs elliptiques dégénérés* (French), Rend. Sem. Mat. Univ. Politec. Torino **Special Issue** (1983), 105–114 (1984). Conference on linear partial and pseudodifferential operators (Torino, 1982). MR745979

[FrLan4] Bruno Franchi and Ermanno Lanconelli, *An embedding theorem for Sobolev spaces related to nonsmooth vector fields and Harnack inequality*, Comm. Partial Differential Equations **9** (1984), no. 13, 1237–1264, DOI 10.1080/03605308408820362. MR764663

[FrLan5] Bruno Franchi and Ermanno Lanconelli, *Une condition géométrique pour l'inégalité de Harnack* (French), J. Math. Pures Appl. (9) **64** (1985), no. 3, 237–256. MR823405

[FrSeSe] Bruno Franchi, Raul Serapioni, and Francesco Serra Cassano, *Approximation and imbedding theorems for weighted Sobolev spaces associated with Lipschitz continuous vector fields* (English, with Italian summary), Boll. Un. Mat. Ital. B (7) **11** (1997), no. 1, 83–117. MR1448000

[GaNh] Nicola Garofalo and Duy-Minh Nhieu, *Lipschitz continuity, global smooth approximations and extension theorems for Sobolev functions in Carnot-Carathéodory spaces*, J. Anal. Math. **74** (1998), 67–97, DOI 10.1007/BF02819446. MR1631642

[GiTr] David Gilbarg and Neil S. Trudinger, *Elliptic partial differential equations of second order*, Springer-Verlag, Berlin-New York, 1977. Grundlehren der Mathematischen Wissenschaften, Vol. 224. MR0473443

[GuLa] Cristian E. Gutiérrez and Ermanno Lanconelli, *Maximum principle, nonhomogeneous Harnack inequality, and Liouville theorems for X-elliptic operators*, Comm. Partial Differential Equations **28** (2003), no. 11-12, 1833–1862, DOI 10.1081/PDE-120025487. MR2015404

[Ho] Lars Hörmander, *Hypoelliptic second order differential equations*, Acta Math. **119** (1967), 147–171, DOI 10.1007/BF02392081. MR0222474

[KoMaRi] Lyudmila Korobenko, Diego Maldonado, and Cristian Rios, *From Sobolev inequality to doubling*, Proc. Amer. Math. Soc. **143** (2015), no. 9, 4017–4028, DOI 10.1090/S0002-9939-2015-12556-7. MR3359590

[KoRiSaSh1] L. Korobenko, C. Rios, E. Sawyer and Ruipeng Shen, *Local boundedness, maximum principles, and continuity of solutions to infinitely degenerate elliptic equations*, `arXiv:1506.09203v5` (2016).

[KoRiSaSh2] L. Korobenko, C. Rios, E. Sawyer and Ruipeng Shen, *Sharp local boundedness, and maximum principle in the infinitely degenerate regime via DeGiorgi iteration*, `arXiv:1506.01630v3` (2017).

[KoRiSaSh3] L. Korobenko, C. Rios, E. Sawyer and Ruipeng Shen, *Continuity of weak solutions to rough infinitely degenerate equations*, `arXiv:1703.00774v2` (2017).
Applications of the Malliavin Calculus II, J. Fac. Sci. Univ. Tokyo

[KuStr] S. Kusuoka and D. Stroock, *Applications of the Malliavin calculus. II*, J. Fac. Sci. Univ. Tokyo Sect. IA Math. **32** (1985), no. 1, 1–76. MR783181

[Lan] E.Lanconelli, *Regolarità holderiana delle soluzioni deboli di certe equazioni ellitiche fortemente degeneri (Hölder regularity of solutions to certain strongly degenerate elliptic equations)* Sem di Analisi Matematica, Ist. Mat. di Bologna (a.a.1981-1982), IX.1-IX.27

[Mor] Yoshinori Morimoto, *Nonhypoellipticity for degenerate elliptic operators*, Publ. Res. Inst. Math. Sci. **22** (1986), no. 1, 25–30, DOI 10.2977/prims/1195178369. MR834345
24 (1971), 727-740.

[Mos] J. Moser, *On a pointwise estimate for parabolic differential equations*, Comm. Pure Appl. Math. **24** (1971), 727–740, DOI 10.1002/cpa.3160240507. MR0288405

[RSaW] Cristian Rios, Eric T. Sawyer, and Richard L. Wheeden, *A higher-dimensional partial Legendre transform, and regularity of degenerate Monge-Ampère equations*, Adv. Math. **193** (2005), no. 2, 373–415, DOI 10.1016/j.aim.2004.05.009. MR2137289

[RSaW1] Cristian Rios, Eric T. Sawyer, and Richard L. Wheeden, *A priori estimates for infinitely degenerate quasilinear equations*, Differential Integral Equations **21** (2008), no. 1-2, 131–200. MR2479665
Hypoellipticity for Infinitely Degenerate Quasilinear Equations and the

[RSaW2] Cristian Rios, Eric T. Sawyer, and Richard L. Wheeden, *Hypoellipticity for infinitely degenerate quasilinear equations and the Dirichlet problem*, J. Anal. Math. **119** (2013), 1–62, DOI 10.1007/s11854-013-0001-6. MR3043146
Regularity of degenerate Monge-Ampère and prescribed Gaussian curvature

[SaW] Eric T. Sawyer and Richard L. Wheeden, *Regularity of degenerate Monge-Ampère and prescribed Gaussian curvature equations in two dimensions*, Potential Anal. **24** (2006), no. 3, 267–301, DOI 10.1007/s11118-005-0915-4. MR2217954
A priori estimates for quasilinear equations related to the Monge-Ampère

[SaW2] Eric T. Sawyer and Richard L. Wheeden, *A priori estimates for quasilinear equations related to the Monge-Ampère equation in two dimensions*, J. Anal. Math. **97** (2005), 257–316, DOI 10.1007/BF02807409. MR2274980

[SaW3] Eric T. Sawyer and Richard L. Wheeden, *Degenerate Sobolev spaces and regularity of subelliptic equations*, Trans. Amer. Math. Soc. **362** (2010), no. 4, 1869–1906, DOI 10.1090/S0002-9947-09-04756-4. MR2574880

[SaWh4] Eric T. Sawyer and Richard L. Wheeden, *Hölder continuity of weak solutions to subelliptic equations with rough coefficients*, Mem. Amer. Math. Soc. **180** (2006), no. 847, x+157, DOI 10.1090/memo/0847. MR2204824

Editorial Information

To be published in the *Memoirs*, a paper must be correct, new, nontrivial, and significant. Further, it must be well written and of interest to a substantial number of mathematicians. Piecemeal results, such as an inconclusive step toward an unproved major theorem or a minor variation on a known result, are in general not acceptable for publication.

Papers appearing in *Memoirs* are generally at least 80 and not more than 200 published pages in length. Papers less than 80 or more than 200 published pages require the approval of the Managing Editor of the Transactions/Memoirs Editorial Board. Published pages are the same size as those generated in the style files provided for $\mathcal{A}\mathcal{M}\mathcal{S}$-LaTeX or $\mathcal{A}\mathcal{M}\mathcal{S}$-TeX.

Information on the backlog for this journal can be found on the AMS website starting from `http://www.ams.org/memo`.

A Consent to Publish is required before we can begin processing your paper. After a paper is accepted for publication, the Providence office will send a Consent to Publish and Copyright Agreement to all authors of the paper. By submitting a paper to the *Memoirs*, authors certify that the results have not been submitted to nor are they under consideration for publication by another journal, conference proceedings, or similar publication.

Information for Authors

Memoirs is an author-prepared publication. Once formatted for print and on-line publication, articles will be published as is with the addition of AMS-prepared frontmatter and backmatter. Articles are not copyedited; however, confirmation copy will be sent to the authors.

Initial submission. The AMS uses Centralized Manuscript Processing for initial submissions. Authors should submit a PDF file using the Initial Manuscript Submission form found at `www.ams.org/submission/memo`, or send one copy of the manuscript to the following address: Centralized Manuscript Processing, MEMOIRS OF THE AMS, 201 Charles Street, Providence, RI 02904-2294 USA. If a paper copy is being forwarded to the AMS, indicate that it is for *Memoirs* and include the name of the corresponding author, contact information such as email address or mailing address, and the name of an appropriate Editor to review the paper (see the list of Editors below).

The paper must contain a *descriptive title* and an *abstract* that summarizes the article in language suitable for workers in the general field (algebra, analysis, etc.). The *descriptive title* should be short, but informative; useless or vague phrases such as "some remarks about" or "concerning" should be avoided. The *abstract* should be at least one complete sentence, and at most 300 words. Included with the footnotes to the paper should be the 2010 *Mathematics Subject Classification* representing the primary and secondary subjects of the article. The classifications are accessible from `www.ams.org/msc/`. The Mathematics Subject Classification footnote may be followed by a list of *key words and phrases* describing the subject matter of the article and taken from it. Journal abbreviations used in bibliographies are listed in the latest *Mathematical Reviews* annual index. The series abbreviations are also accessible from `www.ams.org/msnhtml/serials.pdf`. To help in preparing and verifying references, the AMS offers MR Lookup, a Reference Tool for Linking, at `www.ams.org/mrlookup/`.

Electronically prepared manuscripts. The AMS encourages electronically prepared manuscripts, with a strong preference for $\mathcal{A}\mathcal{M}\mathcal{S}$-LaTeX. To this end, the Society has prepared $\mathcal{A}\mathcal{M}\mathcal{S}$-LaTeX author packages for each AMS publication. Author packages include instructions for preparing electronic manuscripts, samples, and a style file that generates the particular design specifications of that publication series. Though $\mathcal{A}\mathcal{M}\mathcal{S}$-LaTeX is the highly preferred format of TeX, author packages are also available in $\mathcal{A}\mathcal{M}\mathcal{S}$-TeX.

Authors may retrieve an author package for *Memoirs of the AMS* from `www.ams.org/journals/memo/memoauthorpac.html`. The *AMS Author Handbook* is available in PDF format from the author package link. The author package can also be obtained free of charge by sending email to **tech-support@ams.org** or from the Publication Division, American Mathematical Society, 201 Charles St., Providence, RI 02904-2294, USA. When requesting an author package, please specify $\mathcal{A}_{\mathcal{M}}\mathcal{S}$-LaTeX or $\mathcal{A}_{\mathcal{M}}\mathcal{S}$-TeX and the publication in which your paper will appear. Please be sure to include your complete mailing address.

After acceptance. The source files for the final version of the electronic manuscript should be sent to the Providence office immediately after the paper has been accepted for publication. The author should also submit a PDF of the final version of the paper to the editor, who will forward a copy to the Providence office.

Accepted electronically prepared files can be submitted via the web at **www.ams.org/submit-book-journal/**, sent via FTP, or sent on CD to the Electronic Prepress Department, American Mathematical Society, 201 Charles Street, Providence, RI 02904-2294 USA. TeX source files and graphic files can be transferred over the Internet by FTP to the Internet node **ftp.ams.org** (130.44.1.100). When sending a manuscript electronically via CD, please be sure to include a message indicating that the paper is for the *Memoirs*.

Electronic graphics. Comprehensive instructions on preparing graphics are available at www.ams.org/authors/journals.html. A few of the major requirements are given here.

Submit files for graphics as EPS (Encapsulated PostScript) files. This includes graphics originated via a graphics application as well as scanned photographs or other computer-generated images. If this is not possible, TIFF files are acceptable as long as they can be opened in Adobe Photoshop or Illustrator.

Authors using graphics packages for the creation of electronic art should also avoid the use of any lines thinner than 0.5 points in width. Many graphics packages allow the user to specify a "hairline" for a very thin line. Hairlines often look acceptable when proofed on a typical laser printer. However, when produced on a high-resolution laser imagesetter, hairlines become nearly invisible and will be lost entirely in the final printing process.

Screens should be set to values between 15% and 85%. Screens which fall outside of this range are too light or too dark to print correctly. Variations of screens within a graphic should be no less than 10%.

Any graphics created in color will be rendered in grayscale for the printed version unless color printing is authorized by the Managing Editor and the Publisher. In general, color graphics will appear in color in the online version.

Inquiries. Any inquiries concerning a paper that has been accepted for publication should be sent to memo-query@ams.org or directly to the Electronic Prepress Department, American Mathematical Society, 201 Charles St., Providence, RI 02904-2294 USA.

Editors

Selected Published Titles in This Series

1303 **Zhi Qi,** Theory of Fundamental Bessel Functions of High Rank, 2020

1302 **Paul M. N. Feehan and Manousos Maridakis,** Łojasiewicz-Simon Gradient Inequalities for Coupled Yang-Mills Energy Functionals, 2020

1301 **Joachim Krieger,** On Stability of Type II Blow Up for the Critical Nonlinear Wave Equation in $\mathbb{R}^{3+1}$, 2020

1300 **Camille Male,** Traffic Distributions and Independence: Permutation Invariant Random Matrices and the Three Notions of Independence, 2020

1299 **David M. J. Calderbank, Michael G. Eastwood, Vladimir S. Matveev, and Katharina Neusser,** C-Projective Geometry, 2020

1298 **Sy David Friedman and David Schrittesser,** Projective Measure Without Projective Baire, 2020

1297 **Jonathan Gantner,** Operator Theory on One-Sided Quaternion Linear Spaces: Intrinsic S-Functional Calculus and Spectral Operators, 2020

1296 **Christophe Cornut,** Filtrations and Buildings, 2020

1295 **Lisa Berger, Chris Hall, René Pannekoek, Jennifer Park, Rachel Pries, Shahed Sharif, Alice Silverberg, and Douglas Ulmer,** Explicit Arithmetic of Jacobians of Generalized Legendre Curves Over Global Function Fields, 2020

1294 **Jacob Bedrossian, Pierre Germain, and Nader Masmoudi,** Dynamics Near the Subcritical Transition of the 3D Couette Flow I: Below Threshold Case, 2020

1293 **Benjamin Jaye, Fedor Nazarov, Maria Carmen Reguera, and Xavier Tolsa,** The Riesz Transform of Codimension Smaller Than One and the Wolff Energy, 2020

1292 **Angel Castro, Diego Córdoba, and Javier Gómez-Serrano,** Global Smooth Solutions for the Inviscid SQG Equation, 2020

1291 **Vasilis Chousionis, Jeremy Tyson, and Mariusz Urbanski,** Conformal Graph Directed Markov Systems on Carnot Groups, 2020

1290 **Harold Rosenberg and Graham Smith,** Degree Theory of Immersed Hypersurfaces, 2020

1289 **Pavel M. Bleher and Guilherme L. F. Silva,** The Mother Body Phase Transition in the Normal Matrix Model, 2020

1288 **Alexander Blokh, Lex Oversteegen, Ross Ptacek, and Vladlen Timorin,** Laminational Models for Some Spaces of Polynomials of Any Degree, 2020

1287 **Mike Hochman,** On Self-Similar Sets with Overlaps and Inverse Theorems for Entropy in $\mathbb{R}^d$, 2020

1286 **Andrew J. Blumberg and Michael A. Mandell,** Localization for $THH(ku)$ and the Topological Hochschild and Cyclic Homology of Waldhausen Categories, 2020

1285 **Zhaobing Fan, Chun-Ju Lai, Yiqiang Li, Li Luo, and Weiqiang Wang,** Affine Flag Varieties and Quantum Symmetric Pairs, 2020

1284 **Rodney G. Downey, Keng Meng Ng, and Reed Solomon,** Minimal Weak Truth Table Degrees and Computably Enumerable Turing Degrees, 2020

1283 **Antonio Alarcón, Franc Forstnerič, and Francisco J. López,** New Complex Analytic Methods in the Study of Non-Orientable Minimal Surfaces in $\mathbb{R}^n$, 2020

1282 **David Carchedi,** Higher Orbifolds and Deligne-Mumford Stacks as Structured Infinity-Topoi, 2020

1281 **S. V. Ivanov,** The Bounded and Precise Word Problems for Presentations of Groups, 2020

1280 **Michael Handel and Lee Mosher,** Subgroup Decomposition in Out(F_n), 2020

1279 **Cristian Gavrus and Sung-Jin Oh,** Global Well-Posedness of High Dimensional Maxwell–Dirac for Small Critical Data, 2020

For a complete list of titles in this series, visit the
AMS Bookstore at **www.ams.org/bookstore/memoseries/**.